MULTILEVEL STRUCTURAL EQUATION MODELING

Quantitative Applications in the Social Sciences

A SAGE PUBLICATIONS SERIES

1. Analysis of Variance, 2nd Edition *Iversen/Norpoth*
2. Operations Research Methods *Nagel/Neef*
3. Causal Modeling, 2nd Edition *Asher*
4. Tests of Significance *Henkel*
5. Cohort Analysis, 2nd Edition *Glenn*
6. Canonical Analysis and Factor Comparison *Levine*
7. Analysis of Nominal Data, 2nd Edition *Reynolds*
8. Analysis of Ordinal Data *Hildebrand/Laing/Rosenthal*
9. Time Series Analysis, 2nd Edition *Ostrom*
10. Ecological Inference *Langbein/Lichtman*
11. Multidimensional Scaling *Kruskal/Wish*
12. Analysis of Covariance *Wildt/Ahtola*
13. Introduction to Factor Analysis *Kim/Mueller*
14. Factor Analysis *Kim/Mueller*
15. Multiple Indicators *Sullivan/Feldman*
16. Exploratory Data Analysis *Hartwig/Dearing*
17. Reliability and Validity Assessment *Carmines/Zeller*
18. Analyzing Panel Data *Markus*
19. Discriminant Analysis *Klecka*
20. Log-Linear Models *Knoke/Burke*
21. Interrupted Time Series Analysis *McDowall/McCleary/Meidinger/Hay*
22. Applied Regression, 2nd Edition *Lewis-Beck/Lewis-Beck*
23. Research Designs *Spector*
24. Unidimensional Scaling *McIver/Carmines*
25. Magnitude Scaling *Lodge*
26. Multiattribute Evaluation *Edwards/Newman*
27. Dynamic Modeling *Huckfeldt/Kohfeld/Likens*
28. Network Analysis *Knoke/Kuklinski*
29. Interpreting and Using Regression *Achen*
30. Test Item Bias *Osterlind*
31. Mobility Tables *Hout*
32. Measures of Association *Liebetrau*
33. Confirmatory Factor Analysis *Long*
34. Covariance Structure Models *Long*
35. Introduction to Survey Sampling *Kalton*
36. Achievement Testing *Bejar*
37. Nonrecursive Causal Models *Berry*
38. Matrix Algebra *Namboodiri*
39. Introduction to Applied Demography *Rives/Serow*
40. Microcomputer Methods for Social Scientists, 2nd Edition *Schrodt*
41. Game Theory *Zagare*
42. Using Published Data *Jacob*
43. Bayesian Statistical Inference *Iversen*
44. Cluster Analysis *Aldenderfer/Blashfield*
45. Linear Probability, Logit, and Probit Models *Aldrich/Nelson*
46. Event History and Survival Analysis, 2nd Edition *Allison*
47. Canonical Correlation Analysis *Thompson*
48. Models for Innovation Diffusion *Mahajan/Peterson*
49. Basic Content Analysis, 2nd Edition *Weber*
50. Multiple Regression in Practice *Berry/Feldman*
51. Stochastic Parameter Regression Models *Newbold/Bos*
52. Using Microcomputers in Research *Madron/Tate/Brookshire*
53. Secondary Analysis of Survey Data *Kiecolt/Nathan*
54. Multivariate Analysis of Variance *Bray/Maxwell*
55. The Logic of Causal Order *Davis*
56. Introduction to Linear Goal Programming *Ignizio*
57. Understanding Regression Analysis, 2nd Edition *Schroeder/Sjoquist/Stephan*
58. Randomized Response and Related Methods, 2nd Edition *Fox/Tracy*
59. Meta-Analysis *Wolf*
60. Linear Programming *Feiring*
61. Multiple Comparisons *Klockars/Sax*
62. Information Theory *Krippendorff*
63. Survey Questions *Converse/Presser*
64. Latent Class Analysis *McCutcheon*
65. Three-Way Scaling and Clustering *Arabie/Carroll/DeSarbo*
66. Q Methodology, 2nd Edition *McKeown/Thomas*
67. Analyzing Decision Making *Louviere*
68. Rasch Models for Measurement *Andrich*
69. Principal Components Analysis *Dunteman*
70. Pooled Time Series Analysis *Sayrs*
71. Analyzing Complex Survey Data, 2nd Edition *Lee/Forthofer*
72. Interaction Effects in Multiple Regression, 2nd Edition *Jaccard/Turrisi*
73. Understanding Significance Testing *Mohr*
74. Experimental Design and Analysis *Brown/Melamed*
75. Metric Scaling *Weller/Romney*
76. Longitudinal Research, 2nd Edition *Menard*
77. Expert Systems *Benfer/Brent/Furbee*
78. Data Theory and Dimensional Analysis *Jacoby*
79. Regression Diagnostics *Fox*
80. Computer-Assisted Interviewing *Saris*
81. Contextual Analysis *Iversen*
82. Summated Rating Scale Construction *Spector*
83. Central Tendency and Variability *Weisberg*
84. ANOVA: Repeated Measures *Girden*
85. Processing Data *Bourque/Clark*
86. Logit Modeling *DeMaris*
87. Analytic Mapping and Geographic Databases *Garson/Biggs*
88. Working With Archival Data *Elder/Pavalko/Clipp*
89. Multiple Comparison Procedures *Toothaker*
90. Nonparametric Statistics *Gibbons*
91. Nonparametric Measures of Association *Gibbons*
92. Understanding Regression Assumptions *Berry*
93. Regression With Dummy Variables *Hardy*
94. Loglinear Models With Latent Variables *Hagenaars*
95. Bootstrapping *Mooney/Duval*
96. Maximum Likelihood Estimation *Eliason*
97. Ordinal Log-Linear Models *Ishii-Kuntz*
98. Random Factors in ANOVA *Jackson/Brashers*

# Quantitative Applications in the Social Sciences

A SAGE PUBLICATIONS SERIES

99. Univariate Tests for Time Series Models *Cromwell/Labys/Terraza*
100. Multivariate Tests for Time Series Models *Cromwell/Hannan/Labys/Terraza*
101. Interpreting Probability Models: Logit, Probit, and Other Generalized Linear Models *Liao*
102. Typologies and Taxonomies *Bailey*
103. Data Analysis: An Introduction *Lewis-Beck*
104. Multiple Attribute Decision Making *Yoon/Hwang*
105. Causal Analysis With Panel Data *Finkel*
106. Applied Logistic Regression Analysis, 2nd Edition *Menard*
107. Chaos and Catastrophe Theories *Brown*
108. Basic Math for Social Scientists: Concepts *Hagle*
109. Basic Math for Social Scientists: Problems and Solutions *Hagle*
110. Calculus *Iversen*
111. Regression Models: Censored, Sample Selected, or Truncated Data *Breen*
112. Tree Models of Similarity and Association *Corter*
113. Computational Modeling *Taber/Timpone*
114. LISREL Approaches to Interaction Effects in Multiple Regression *Jaccard/Wan*
115. Analyzing Repeated Surveys *Firebaugh*
116. Monte Carlo Simulation *Mooney*
117. Statistical Graphics for Univariate and Bivariate Data *Jacoby*
118. Interaction Effects in Factorial Analysis of Variance *Jaccard*
119. Odds Ratios in the Analysis of Contingency Tables *Rudas*
120. Statistical Graphics for Visualizing Multivariate Data *Jacoby*
121. Applied Correspondence Analysis *Clausen*
122. Game Theory Topics *Fink/Gates/Humes*
123. Social Choice: Theory and Research *Johnson*
124. Neural Networks *Abdi/Valentin/Edelman*
125. Relating Statistics and Experimental Design: An Introduction *Levin*
126. Latent Class Scaling Analysis *Dayton*
127. Sorting Data: Collection and Analysis *Coxon*
128. Analyzing Documentary Accounts *Hodson*
129. Effect Size for ANOVA Designs *Cortina/Nouri*
130. Nonparametric Simple Regression: Smoothing Scatterplots *Fox*
131. Multiple and Generalized Nonparametric Regression *Fox*
132. Logistic Regression: A Primer *Pampel*
133. Translating Questionnaires and Other Research Instruments: Problems and Solutions *Behling/Law*
134. Generalized Linear Models: A Unified Approach, 2nd Edition *Gill/Torres*
135. Interaction Effects in Logistic Regression *Jaccard*
136. Missing Data *Allison*
137. Spline Regression Models *Marsh/Cormier*
138. Logit and Probit: Ordered and Multinomial Models *Borooah*
139. Correlation: Parametric and Nonparametric Measures *Chen/Popovich*
140. Confidence Intervals *Smithson*
141. Internet Data Collection *Best/Krueger*
142. Probability Theory *Rudas*
143. Multilevel Modeling *Luke*
144. Polytomous Item Response Theory Models *Ostini/Nering*
145. An Introduction to Generalized Linear Models *Dunteman/Ho*
146. Logistic Regression Models for Ordinal Response Variables *O'Connell*
147. Fuzzy Set Theory: Applications in the Social Sciences *Smithson/Verkuilen*
148. Multiple Time Series Models *Brandt/Williams*
149. Quantile Regression *Hao/Naiman*
150. Differential Equations: A Modeling Approach *Brown*
151. Graph Algebra: Mathematical Modeling With a Systems Approach *Brown*
152. Modern Methods for Robust Regression *Andersen*
153. Agent-Based Models *Gilbert*
154. Social Network Analysis, 2nd Edition *Knoke/Yang*
155. Spatial Regression Models, 2nd Edition *Ward/Gleditsch*
156. Mediation Analysis *Iacobucci*
157. Latent Growth Curve Modeling *Preacher/Wichman/MacCallum/Briggs*
158. Introduction to the Comparative Method With Boolean Algebra *Caramani*
159. A Mathematical Primer for Social Statistics *Fox*
160. Fixed Effects Regression Models *Allison*
161. Differential Item Functioning, 2nd Edition *Osterlind/Everson*
162. Quantitative Narrative Analysis *Franzosi*
163. Multiple Correspondence Analysis *LeRoux/Rouanet*
164. Association Models *Wong*
165. Fractal Analysis *Brown/Liebovitch*
166. Assessing Inequality *Hao/Naiman*
167. Graphical Models and the Multigraph Representation for Categorical Data *Khamis*
168. Nonrecursive Models *Paxton/Hipp/Marquart-Pyatt*
169. Ordinal Item Response Theory *Van Schuur*
170. Multivariate General Linear Models *Haase*
171. Methods of Randomization in Experimental Design *Alferes*
172. Heteroskedasticity in Regression *Kaufman*
173. An Introduction to Exponential Random Graph Modeling *Harris*
174. Introduction to Time Series Analysis *Pickup*
175. Factorial Survey Experiments *Auspurg/Hinz*
176. Introduction to Power Analysis: Two-Group Studies *Hedberg*
177. Linear Regression: A Mathematical Introduction *Gujarati*
178. Propensity Score Methods and Applications *Bai/Clark*
179. Multilevel Structural Equation Modeling *Silva/Bosancianu/Littvay*

Sara Miller McCune founded SAGE Publishing in 1965 to support the dissemination of usable knowledge and educate a global community. SAGE publishes more than 1000 journals and over 800 new books each year, spanning a wide range of subject areas. Our growing selection of library products includes archives, data, case studies and video. SAGE remains majority owned by our founder and after her lifetime will become owned by a charitable trust that secures the company's continued independence.

Los Angeles | London | New Delhi | Singapore | Washington DC | Melbourne

MULTILEVEL STRUCTURAL EQUATION MODELING

Bruno Castanho Silva
University of Cologne

Constantin Manuel Bosancianu
WZB Berlin Social Science Center

Levente Littvay
Central European University

Los Angeles | London | New Delhi
Singapore | Washington DC | Melbourne

Los Angeles | London | New Delhi
Singapore | Washington DC | Melbourne

FOR INFORMATION:

SAGE Publications, Inc.
2455 Teller Road
Thousand Oaks, California 91320
E-mail: order@sagepub.com

SAGE Publications Ltd.
1 Oliver's Yard
55 City Road
London EC1Y 1SP
United Kingdom

SAGE Publications India Pvt. Ltd.
B 1/I 1 Mohan Cooperative Industrial Area
Mathura Road, New Delhi 110 044
India

SAGE Publications Asia-Pacific Pte. Ltd.
18 Cross Street #10-10/11/12
China Square Central
Singapore 048423

Acquisitions Editor: Helen Salmon
Editorial Assistant: Megan O'Heffernan
Production Editor: Rebecca Lee
Copy Editor: Gillian Dickens
Typesetter: Integra
Proofreader: Wendy Jo Dymond
Indexer: Marilyn Augst
Cover Designer: Candice Harman
Marketing Manager: Shari Countryman

Copyright © 2020 by SAGE Publications, Inc.

All rights reserved. Except as permitted by U.S. copyright law, no part of this work may be reproduced or distributed in any form or by any means, or stored in a database or retrieval system, without permission in writing from the publisher.

All third party trademarks referenced or depicted herein are included solely for the purpose of illustration and are the property of their respective owners. Reference to these trademarks in no way indicates any relationship with, or endorsement by, the trademark owner.

Library of Congress Cataloging-in-Publication Data

Names: Silva, Bruno Castanho, author. | Bosancianu, Constantin Manuel, author. | Littvay, Levente, author.

Title: Multilevel structural equation modeling / Bruno Castanho Silva (University of Cologne), Constantin Manuel Bosancianu (Wissenschaftszentrum Berlin für Sozialforschung), Levente Littvay (Central European University).

Description: Thousand Oaks, California : SAGE Publications, Inc., 2019. | Includes bibliographical references and index.

Identifiers: LCCN 2018057333 | ISBN 978-1-5443-2305-3 (pbk. : alk. paper)

Subjects: LCSH: Path analysis (Statistics) | Structural equation modeling. | Multilevel models (Statistics) | Regression analysis.

Classification: LCC QA278.3 .S55 2019 | DDC 519.5/3–dc23 LC record available at https://lccn.loc.gov/2018057333

19 20 21 22 23 10 9 8 7 6 5 4 3 2 1

CONTENTS

Series Editor's Introduction	xi
About the Authors	xiii
Acknowledgments	xv
1. Introduction	**1**
About the Book and MSEM	1
Quick Review of Structural Equation Models	2
Quick Review of Multilevel Models	15
Introduction to MSEM and Its Notation	20
Estimation and Model Fit	29
Scope of the Book and Online Materials	30
2. Multilevel Path Models	**31**
Multilevel Regression Example	34
Random Intercepts Model	36
Random Slopes Model	41
Comparison of Random Intercepts and Random Slopes Models	45
Mediation and Moderation	45
Summary	52
3. Multilevel Factor Models	**54**
Confirmatory Factor Analysis in Multiple Groups	57
Two-Level CFA	58
Random Latent Variable Intercepts	66
Multilevel CFA With Random Loadings	69
Summary	76
4. Multilevel Structural Equation Models	**78**
Bringing Factor and Path Models Together	78
Random Intercept of Observed Outcome	80
Multilevel Latent Covariate Model	85
Structural Models With Between-Level Latent Variables	88
Random Slopes MSEM	98
Summary	104
5. Conclusion	**106**
References	**116**
Index	**123**

LIST OF FIGURES

1.1	SEM Graphical Notation	4
1.2	Example Path Model With Latent Variables	4
1.3	Path Model in Graphical Form	5
1.4	MLM Graphical Notation	15
1.5	Multilevel Path Model Translated to Graphical Form	22
1.6	Multilevel Path Model With Random Intercepts and Slopes Translated to Graphical Form	23
1.7	Full SEM Model With Random Intercepts and Random Slopes Translated to Graphical Form	25
2.1	Standard Multilevel Regression Model Specification (unstandardized results reported)	35
2.2	Standard Path Model	37
2.3	Multilevel Path Model With Random Intercepts (unstandardized results reported)	39
2.4	Bivariate Relationships Between Education, Self-Expression Values, and Age (individual country fit lines)	41
2.5	Multilevel Path Model With Random Intercepts and Slopes (unstandardized coefficients)	43
2.6	Basic Framework for Statistical Mediation and Moderation	46
2.7	Moderated Mediation Effect in the Slopes Model	49
3.1	Basic CFA Model of Digital Devices Use	56
3.2	Two-Level CFA Model of Digital Devices Usage	62
3.3	Two-Level CFA Model of Digital Devices Use	63
3.4	Two-Level CFA Model of Digital Devices Use With a Unidimensional Level 2 Model	65
3.5	Two-Level CFA Model of Digital Devices Use With a Unidimensional Level 2 Model and Random Factor Loadings	74
4.1	Two-Level Model of Perception of Own Skills With a Random Intercept	81
4.2	Two-Level Model of Perception of Own Skills With a Random Intercept	84
4.3	Latent Contextual Covariate Model	87

4.4	Random Intercepts Model With Between-Level Latent Variables	90
4.5	Random Intercepts Model With Between-Level Latent Variables (unstandardized results)	95
4.6	Random Intercepts Model With Random Mean Latent Variables	97
4.7	Random Slopes Model With Random Mean Latent Variables	99
4.8	Random Slopes Model (unstandardized results)	102

SERIES EDITOR'S INTRODUCTION

I am pleased to introduce *Multilevel Structural Equation Modeling*, by Bruno Castanho Silva, Constantin Manuel Cosanciau, and Levente Littvay. Multilevel structural equation models (MSEMs) combine the study of relationships between variables measured with error central to structural equation modeling (SEM) with an interest in macro-micro relationships central to multilevel models (MLM). This volume is well-organized with a clear progression of topics, starting with SEMs of observed variables, proceeding to confirmatory factor analysis (CFA), and then the full model, adding a multilevel component to each along the way. In each chapter, the authors proceed systematically from simpler to more complex model specifications, using examples to illustrate each step. Readers can practice by replicating these examples using materials available in the online appendix.

An innovation of the volume is the notation. SEMs and MLMs each have their own conventions; the authors blend them. They maintain notation used in standard multilevel texts but introduce superscripts to keep track of the outcome variable associated with particular coefficients. This approach makes it possible to write MSEM models as a series of equations, making the volume broadly accessible, even to readers not well-versed in matrix algebra. To further reinforce the reader understanding, they show most models both as a set of equations and also as graphical presentations that build on SEM traditions.

In terms of the preparation needed, readers having experience with structural equations models (SEM) or multilevel models (MLM) are in the best position to benefit from this volume. Chapter 1 provides a very helpful review of both, and then shows how the model and notation can be organized into a single framework for MSEM. Chapter 2 introduces multilevel path models, considering both the random intercept model and the random slopes model. It uses World Values Survey data, Wave 4, from 55 countries to explore the individual-level and country-level factors associated with high self-expression values (importance of civic activism, subjective well-being, tolerance and trust, personal autonomy and choice). Multilevel factor models are the focus of Chapter 3. This chapter uses data from the 2015 wave of the Program for International Student Assessment (PISA) on the use of digital devices in the Dominican Republic to build a two-level CFA, starting with a comparison of the multilevel CFA to the multiple group CFA, then building in random latent variable intercepts and finishing with a multilevel CFA with random loadings. It also includes a useful discussion of measurement invariance. Chapter 4 merges the subject matter of chapters 2 and 3 together into the full multilevel structural equation model (MSEM). The example for this chapter, based on the 2004 Workplace Employment Relations Survey teaching dataset, explores whether employees consider themselves to be under- or over-qualified for their jobs, as affected by their perception of how demanding their jobs are, how responsive their managers, their pay, and—at the company level—number of employees. Chapter 5 concludes the text by addressing some advanced topics such as categorical dependent variables, sampling weights, and missing data, pointing to references where the interested reader can learn more and providing advice on how to approach the technical literature.

Multilevel structural equation models can be quite complex. Indeed, as the authors say, the complexity of the models to be investigated is only limited by the imagination of the investigators (and of course, the data, software, etc.). Given this, readers will especially appreciate this hands-on introduction and the lengths to which the authors have gone to make the material accessible to researchers from a variety of backgrounds.

—Barbara Entwisle
Series Editor

ABOUT THE AUTHORS

Bruno Castanho Silva is a postdoctoral researcher at the Cologne Center for Comparative Politics (CCCP), University of Cologne. Bruno received his PhD from the Department of Political Science at Central European University and teaches introductory and advanced quantitative methods courses, including multilevel structural equation modeling and machine learning, at the European Consortium for Political Research Methods Schools. His methodological interests are on applications of structural equation models for scale development and causal analysis, as well as statistical methods of causal inference with observational and experimental data.

Constantin Manuel Bosancianu is a postdoctoral researcher at the WZB Berlin Social Science Center, in the Institutions and Political Inequality research unit. He received his PhD from the Department of Political Science at Central European University in Budapest, Hungary, and has been an instructor for multiple statistics courses and workshops at the European Consortium for Political Research Methods Schools, at the Universities of Heidelberg, Giessen, and Zagreb, as well as at the Institute of Sociology of the Czech Academy of Sciences. Manuel's methodological focus is on practical applications of multilevel models, Bayesian analysis, and the analysis of time-series cross-sectional data sets.

Levente Littvay is an associate professor at Central European University's Department of Political Science. He is a recipient of the institution's Distinguished Teaching Award for graduate courses in research methods and applied statistics with a topical emphasis on political psychology, experiments, and American politics. He received an MA and a PhD in Political Science and an MS in Survey Research and Methodology from the University of Nebraska–Lincoln, has taught numerous methods workshops, and is an academic co-convenor of the European Consortium for Political Research Methods Schools. His research interests include populism, political socialization, and biological explanations of social and political attitudes and behaviors. He often works as a methodologist with medical researchers and policy analysts, co-runs the Hungarian Twin Registry, is an associate editor for social sciences of *Twin Research and Human Genetics*, and publishes in both social science and medical journals.

ACKNOWLEDGMENTS

Throughout the three years in which this monograph was conceived, drafted, and refined, we have benefited from the kind advice and guidance of a great many of our colleagues. Heartfelt thanks go out to M. Murat Ardag, Nemanja Batrićević, Alexander Bor, Amélie Godefroidt, Jochen Mayerl, Martin Mölder, Ulrich Schroeders, and Federico Vegetti for offering feedback at different stages of the project. Without their consistent help, the difficult content we cover would have been even more impenetrable. We also extend our thanks to participants in the workshops and courses on multilevel structural equation models that we have led during this time. These include the October 2017 MSEM workshop at the University of Bamberg organized by Thomas Saalfeld, especially to Sebastian Jungkunz; the October 2018 MSEM workshop at the Meth-Lab of the Katholieke Universiteit Leuven organized by Amélie Godefroidt and Lala Muradova; two courses in multilevel structural equation modeling (MSEM) taught at the 2016 and 2017 editions of the European Consortium for Political Research (ECPR) Summer School in Methods and Techniques at Central European University in Budapest, Hungary; and, finally, a course in advanced structural equation modeling (SEM) taught at the 2017 edition of the same Summer School. Their questions have repeatedly challenged our thinking, in addition to highlighting areas where our explanations could be clearer.

Last but not least, we owe a debt of gratitude to Yves Rosseel for creating the lavaan scripts for some of the models in this book and Linda Muthén of the *Mplus* team for her speedy response to our questions. Whatever errors have persisted in the book are entirely our own and represent but a small sample of what could have been had the previously mentioned colleagues not generously shared their time, thoughts, and expertise with us.

Our work on this manuscript has been supported in many other ways as well, chief among them being the support staff at the various methods schools and workshops where we have tested our ideas. We wish to thank Miriam Schneider and Dagmar Riess at the University of Bamberg, Anna Foley and Becky Plant from ECPR's Central Services, and the local organizing team in Budapest, especially Carsten Q. Schneider and Robert Sata. We are also grateful for the opportunity to teach these courses to Derek Beach and Benoît Rihoux, who join Levente Littvay as the academic conveners of the ECPR Methods Schools.

Separately, we wish to extend our gratitude to additional valued colleagues. Manuel Bosancianu wishes to thank Zoltán Fazekas for answering many multilevel modeling (MLM) questions over the years, as well as Macartan Humphreys for his feedback and support. Levente Littvay wishes to thank Elmar Schlüter, Bengt Muthén, and Geoffrey Hubona for the MSEM-specific inspiration and, more generally, all the people who handed him QASS series books throughout his studies: Kevin Smith (v22, v79), Brian Humes (v122), Julia McQuillian (v143), Craig Enders (v136), Jim Bovaird (v95, v116, v144, and probably more), and to the loving memory of Allan McCutcheon (v64, but also v126, v119), who started his post-PhD career even before his PhD started by putting him in touch with Tamás Rudas (who wrote v119 and v142). These people, along with Les Hayduk and Mike Neale, shaped his methodological thinking the most and gave him the tools to pass knowledge on to the next generation of outstanding scholars, such as his coauthors and many of the people thanked above for their help and support. This book is dedicated to them as it is very much also their contribution (although he is fairly sure Les and Tamás would not approve much of what is written here, but maybe the final paragraph).

Collectively, we would like to express our gratitude to Barbara Entwisle, Katie Metzler, Megan O'Heffernan, and Helen Salmon from SAGE for their logistic support, advice, and understanding along the entire length of the writing process, as well as to the multiple anonymous reviewers who suggested many invaluable improvements and corrections.

Publisher's Acknowledgments

SAGE would also like to acknowledge the valuable contributions of the following reviewers:

Carl Berning, University of Mainz
Kyungkook Kang, University of Central Florida
Carl L. Palmer, Illinois State University
Ronny Scherer, University of Oslo

Companion Website

A website for the book at http://levente.littvay.hu/msem/ includes replication codes for lavaan for R and *Mplus*, data for all the examples reported, and two LaTeX files to produce all equations and figures from the book.

CHAPTER 1. INTRODUCTION

About the Book and MSEM

Multilevel modeling (MLM) and structural equation modeling (SEM) have become two of the most used techniques in the social sciences. Combining the power of structural equation models with multilevel data, which, until now, are still mostly modeled as one outcome linked to multiple covariates, was inevitable. To date, no comprehensive treatment of the subject exists. What is available is hidden in chapters of SEM and MLM textbooks, or in highly technical treatments of the subject in edited volumes. It is our goal to provide an accessible but comprehensive introduction to multilevel structural equation modeling to social scientists. We assume no specific disciplinary background. Our examples come from political science and/or sociology, education research, and organizational behavior. We use path figures, as SEM scholars are used to them, and develop an equation-based notation that should be familiar to multilevel modelers. While sticking to a multiple equation framework results in models that can become quite big and cumbersome, they are still more accessible than the admittedly more parsimonious matrix algebraic formulation of multilevel structural equation models. We start the book with a review of SEM, MLM, and the development of the notation that is used throughout this volume. Then we devote a chapter each to multilevel path models (2), multilevel confirmatory factor models (3), and multilevel structural models (4).

In each of the chapters, we strive to cover the topics in a concise manner. In the concluding chapter, we point the reader in new and more advanced directions absent from the main chapters of an introductory book. Beyond the restricted case of continuous endogenous variables and a single level of aggregation lie a multitude of empirical situations. The researcher might encounter instances of categorical endogenous variables, as well as additional levels in the data hierarchy. More advanced applications might even require addressing issues of missing information on exogenous or endogenous indicators, as well as incorporating weights in the estimation process. In the context of the series, we simply cannot offer even a basic substantive introduction to these topics, but we provide the reader with suggestions of existing work where guidance might be sought.

Quick Review of Structural Equation Models

SEM emerged to tackle two problems that traditional regression-based models have. To highlight the first problem: What if our research questions go beyond having one outcome and many covariates? What if we want to test a more complex structure of relationships in which we are curious about both direct and indirect effects of X on Y (potentially going through a series of mediators)? SEM's origins may be traced back to biologist Sewell Wright, who developed *path analysis* models in the first half of the 20th century (Schumacker & Lomax, 2004). Path models hypothesize a chain of statistical associations. In its most basic format, X is a covariate for M, and M is a covariate for Y (where X is the exogenous variable as it is not explained by anything, M is the mediator and Y is the outcome, both of which would also be called endogenous variables). The complexity of models tested can grow much beyond that, and this is its first attractiveness: It allows researchers to test an intricate structure of relationships among variables in a single model, in accordance with the kind of elaborate theories that are typical in social sciences.

Technically, SEM refers to a statistical approach to data analysis where multiple structural (i.e., regression) equations are estimated simultaneously to test a set of relationships between variables. Through the pioneering work of Karl Jöreskog (1973), Ward Keesling (1972), and David Wiley (1973), who developed what was first known as the JKW model, SEM was extended to cover what is today its most popular use: causal models with *latent variables*, incorporating *confirmatory factor analysis* (CFA) into path analysis.

Latent variables, also known as factors, solve the second problem associated with the simple regression framework. These are variables that cannot be observed directly by the researcher but are measured by one or multiple observable indicators that, according to theory, capture the latent construct. The common variance of these indicators is taken to represent the latent construct. Several concepts of interest in the social sciences may be seen as latent variables. These models have become popular, for instance, in psychology, where much research is done around topics like depression, aggression, or personality traits, none of which can be directly measured. Psychological constructs are measured by multiple survey items, and the common variation of the items makes up the latent variable. This process eliminates item-specific variance and produces a measurement error–free construct. SEM and path modeling have been incorporated into all of the social sciences and hold an important place in biology, within genetic studies, where some concepts such as heritability are also not directly measurable. In the next pages, we introduce the basic terminology and components of SEM.

Model Specification and Identification

As highlighted above, path models are those with a chain of associations among observed variables: In a minimal format, M is a covariate for Y, while in turn being an outcome for X. It could be written as seen in Equation 1.1.

$$\begin{cases} Y_i = \beta_{01} + \beta_{11} M_i + \varepsilon_{1i} \\ M_i = \beta_{02} + \beta_{12} X_i + \varepsilon_{2i} \end{cases} \quad (1.1)$$

This formulation, though, can grow in complexity to accommodate many more than two outcome variables. In the equation, $i = 1, 2, \ldots, N$ designates individual observations, β_{01} and β_{02} are the intercepts for the endogenous variables in the model, and β_{11} and β_{12} are the slopes for their respective covariates. Finally, ε_{1i} and ε_{2i} are the residuals corresponding to each data point in the sample.

Latent variable models are those in which at least one variable of interest is not directly measured. The most common application is through CFA, in which the variance of the latent construct is estimated based on the covariances of observed indicators. In these models, each indicator is treated as an outcome. For example, being depressed (a latent construct) leads someone to answer a question in a certain way and usually not the other way around. Therefore, CFA models can also be specified as several simultaneous regression equations. When we insert a latent variable into a path model, we get what is called a *full* structural equation model.

Because the number of equations can quickly grow with more complex structures of associations and the addition of latent variables, in most applied research, structural equation models are presented in graphical notation, following the convention in Figure 1.1. We present a didactic example of a model in Figure 1.2 with two latent variables, *feeling of threat* (*THR*) and *xenophobia* (*XEN*). Each is measured by three survey questions (*Q1–Q6*). This is the measurement part of the model. The structural part specifies a path with two relationships: Feeling threatened is linked to xenophobia, which itself is associated with propensity to vote (PTV) for a radical-right party, a continuous measure of party support.

The triangle with a 1 inside indicates this model has a so-called mean structure—intercepts of endogenous variables (squares with straight

Figure 1.1   SEM Graphical Notation

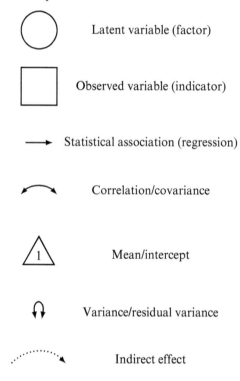

Figure 1.2   Example Path Model With Latent Variables

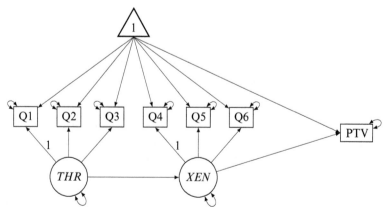

arrows going into them) are estimated much like it is done by default in the regression framework, and means are estimated in addition to the variance–covariance matrix. In fact, the notation of the number 1 in a triangle comes from the inclusion of the vector of 1s in a regression's model matrix. In SEM, including the mean structure in the estimation is becoming the norm—for example, it is necessary for more sophisticated treatments of missing data—although it is still rare to include intercepts in graphs. However, multilevel SEM is a special case where this is quite essential, because higher-level variables can be used to explain the variance of lower-level intercepts. Each arrow from the triangle to a variable denotes the mean of exogenous variables (squares that only serve as the origin of straight arrows but not their targets) or the intercept for endogenous ones.

Curved arrows that go from a variable into itself have different meanings for endogenous and exogenous variables. For all observed endogenous variables, these are the residual variances, also called *errors*. For the exogenous latent variable (THR), it is the estimated factor variance, while for the endogenous latent variable (XEN), it is, once again, the residual variance, which, in the case of latents, is often called a *disturbance*. Straight arrows connecting latent variables to indicators are *factor loadings*. They are a regression coefficient of the indicator regressed on the latent variable.[1] Using this notation, the model introduced in Equation 1.1 could be depicted as presented in Figure 1.3.

If either a curved covariance arrow or a straight arrow is dashed (not presented in Figure 1.3), the implication is that the estimated effect is not statistically significant. We use the following practice for denoting statistical significance in this book: Dashed arrows and an *n.s.* superscript indicate nonsignificant estimates. Solid arrows indicate significant

Figure 1.3 Path Model in Graphical Form

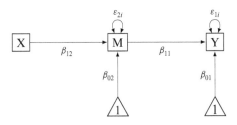

---

[1] We also describe this through simultaneous equation notation later in this chapter.

relationships, alongside stars next to the estimates, with $*p < .05$, $**p < .01$, and $***p < .001$. The two exceptions where we do not use stars, even if technically they should be there, are (1) factor loading estimates where the loadings of reasonably fitting models are always highly significant and (2) results from Bayesian models, where the inferential interpretation is different from frequentist hypothesis tests generalizing to the population. Finally, a dotted curved arrow represents an indirect effect transmission mechanism linking the two variables at either end of the arrow (we borrow the convention from Tinnermann, Geuter, Sprenger, Finsterbusch, & Büchel, 2017).

Identification

In structural equation modeling, it is important to ensure that there is enough empirical information to estimate all the parameters in the model. This is designated by the term *identification*. Much like with other statistical techniques, to arrive at a unique solution, we need to ensure that there are more pieces of known information than unknowns.

Useful known empirical information are the variances and covariances, which form the observed variance-covariance matrix, as well as the vector of observed means, for models with a mean structure. *Degrees of freedom* are the difference between the amount of unique empirical information (entries in the variance-covariance matrix and the means vector) and the number of parameters that are being estimated (i.e., factor loadings, factor variances, errors and disturbances, and regression coefficients). A model is said to be *underidentified* if the degrees of freedom are negative. In this case, it is mathematically possible to estimate parameter values, but they are often not reliable since there are multiple possible solutions. A *just-identified* model is one in which the degrees of freedom is zero. It meets the minimal requirement to be estimated but will always produce a perfect fit to the data, meaning that its goodness of fit cannot be evaluated. It is preferable, therefore, that a model is *overidentified*, or has positive degrees of freedom.

A good heuristic to calculate the number of nonredundant elements in the observed variance-covariance matrix is the formula $p(p + 1)/2$, where p is the number of observed variables in the model (Raykov & Marcoulides, 2000). Given a mean structure, this formula includes an additional term, becoming $p(p + 1)/2 + p$, since there is also a vector of observed means. In the case of Figure 1.2, $p = 7$, so we have 35 nonredundant elements. A total of 22 parameters are estimated in this model: 7 error terms, 1 disturbance, 1 factor variance, 4 factor loadings, 2 regression coefficients, and 7 intercepts. The degrees of freedom are $35 - 22 = 13$, and our model is therefore (over)identified.

It is important to notice, however, that degrees of freedom are a necessary but not sufficient condition for identification. There is also what is referred to as *empirical underidentification* (Kline, 2015). For example, in cases of extreme collinearity, where two variables are highly correlated, in practice the two add only one piece of information. Misspecification, or violating normality and linearity assumptions when using estimation procedures that rely on these, can also lead to empirical underidentification. These issues usually can be detected by data screening.

From Figure 1.2, we see that the model only estimates four of the six total factor loadings. This is on purpose, as in CFA it is necessary to define the metric of each latent variable. This can be done in three ways. The first approach fixes the factor variance(s) (or the residual variances) to a constant, usually 1. The second is fixing the factor loading of one of the factor's indicators (again, usually to 1). The latter approach is far more common and is the default in most SEM software. Depending on whether a factor is an exogenous or endogenous variable, it has, respectively, either a variance or a residual variance. Fixing residual (or unexplained) variances to 1 is not very intuitive as it makes the factor's total variance arbitrarily depend on its covariates' explanatory power. In the example model, there are two factor loadings being estimated for *feeling of threat* and two for *xenophobia*. Fixed factor loadings are denoted by the number "1" next to the arrows in the model. This limits our ability to assess the factor loading's hypothesis test (which is rarely a relevant piece of information in a well-specified model with adequate sample size). Despite the lack of this estimate, through the assessment of standardized estimates for the loadings, it is still possible to check how well the indicator is explained by the latent. Finally, the third approach, most common in developmental psychology, involves fixing the average intercept of indicators to 0 and the average factor loadings to 1 (Little, 2013, p. 93).

Once the model is specified and correctly identified, it is time to estimate and check whether it fits the data.

Estimation

If a model is overidentified, it means that each element of the observed variance-covariance matrix (also called the **S** matrix) can be decomposed into a unique combination of variances, covariances, and parameters (i.e., each element can be defined by a unique equation). By attributing values to the parameters, it is possible to estimate each entry of a model-implied variance-covariance matrix (the Σ matrix). The most common procedure in SEM to compute these parameters is *simultaneous estimation*. It works by deriving all parameters in the model at the same time, under the assumption that the model is correctly specified. The primary approach in this category is maximum likelihood (ML; Eliason, 1993). Conceptually, the goal of ML is to identify a set of population parameters that are most likely to produce the sample at hand (see Enders, 2010, ch. 3). Given a set of parameters, one may calculate the sample likelihood of the data as the sum of individual probabilities (likelihoods) of drawing each data point. When population parameters are unknown, ML estimation works as an iterative process of attributing various sets of values to the parameters and calculating the sample likelihood under each set, until a point when the highest likelihood is found (or the improvement after each iteration becomes too small, below a certain convergence criterion). The parameter values that yield the maximum sample likelihood are the estimated population values. Because likelihoods are very small numbers, sometimes there might be computational problems due to rounding errors. For this reason, as well as the fact that working with sums produced by a logarithm is easier than working with products, the log-likelihood (LL) is used for computing ML estimates. Since the likelihood is a product of densities, each usually bounded by 0 and 1, it is also habitually (but not always) between 0 and 1. In the most common case, then, the values for the logarithm of the likelihood range from $-\infty$ to 0.

In SEM, maximum likelihood estimation consists of iteratively attributing varying values to the parameters and producing a new Σ matrix up to the point when its distance to the S matrix is minimized (Kline, 2015, p. 235). The estimation starts with a random set of initial values for parameters and continuously updates these in a search for the smallest discrepancy between observed and model-implied covariances. Although somewhat computationally intensive, ML estimation is currently the dominant approach with respect to obtaining SEM parameters. It produces global fit measures on the basis of which multiple competing models can be evaluated, and through slightly more complex procedures such as robust maximum likelihood (MLR) or the Bollen-Stine bootstrap (Bollen & Stine, 1992), it can also handle modest deviations from multivariate normality.[2]

---

[2] ML and MLR estimators have equivalents that correct for missing data. These sum the likelihood over the variance-covariance matrices for all of the missing data patterns present in the data. For this reason, they require the raw data as opposed to the variance-covariance matrix. As is also the case in single-level ones, in multilevel models, full-information maximum likelihood (FIML) can give less biased estimates in the presence of missing data, assuming it is missing at random, MAR, or completely at random, MCAR (Enders, 2010; Rubin, 1976).

It is important to note that SEM can be fit with several kinds of estimators, including Bayesian (Kaplan & Depaoli, 2012; B. O. Muthén & Asparouhov, 2012) and least squares–based methods. The latter are usually encountered in *single-equation estimation* approaches, such as two-stage or three-stage least squares (2SLS or 3SLS). Because these procedures usually produce less efficient estimates compared to ML and do not yield any global measures of fit, they have fallen out of favor in recent times. This is not to say that they do not come with advantages. When compared to iterative procedures, such as ML, *single-equation estimation* is less computationally intensive and is generally more robust to specification errors in the model (Kline, 2015, p. 235).

Model Fit

Another main difference between SEM and traditional regression modeling is the concern with model fit. Researchers using SEM may hypothesize and test very complex structures of relationships between variables, so it becomes essential to find reliable ways of testing whether the assumed model is good or not, beyond the usual R^2 of the linear regression world. Good fit denotes that the hypothesized structure of relationships is likely close to the data-generating process, and therefore one can interpret the estimated coefficients as telling us something about the true relationship between those variables. If a model has poor fit, this means that the hypothesized relationships are not in line with the data, and therefore, the estimated coefficients are meaningless and should not be interpreted or given any substantive value.

A model goodness of fit is indicated by the closeness between the observed and the estimated covariance matrices. Over the past 40 years, numerous fit statistics have been proposed in the SEM literature, each with its own strengths and weaknesses. The most common ones are based on the (log-)likelihood, and here we review those that are applicable to MSEM as well. We start out with comparative indices that are most generally applicable to MSEM: the likelihood ratio test, Akaike's information criterion (AIC; Akaike, 1973), and Schwarz's Bayesian information criterion (BIC; Schwarz, 1978).

All of these measures incorporate the *deviance*, a by-product measure of fit based on the likelihood. *Deviance* is a term most commonly used in the multilevel modeling literature, resting on the foundation of maximum likelihood estimation, and as such is also conceptually applicable to SEM. It is computed as $-2LL$, making it an indicator of *badness of fit*: Higher values denote worse-fitting models. While the value of the deviance itself is not very informative, it is the foundation of most fit statistics and has some unique properties that, in addition to AIC and BIC, allow us to effectively use it as a comparative fit statistic. Namely, the difference of two deviance scores has a χ^2 distribution, with degrees of freedom equal to the difference in the number of parameters estimated by the two models. Should the difference in the deviances surpass the critical value from a χ^2 distribution, we can conclude that the more parsimonious model fits the data significantly worse compared to the more complex model.[3] It is important to note, though, that only models using the same cases from the same data can be compared. Two models' likelihoods are only directly comparable with this so-called likelihood ratio test if the models are "nested."[4] Two models are nested if the more parsimonious one of the two can be derived by simply fixing estimates of the more complex one: Those estimates are fixed to be equal to one another or to 0 (hence eliminating the estimation of some of the relationships from the less parsimonious model).

Reader be warned: It is easy to make mistakes when trying to make such comparisons. If we use two different operationalizations of a construct in the models we want to compare, the models are suddenly not nested anymore. The first model has one variable in it, while the second has another variable. There is no avenue to get to the first model solely through the elimination of paths in the second model, and this is why these models are not nested. It is also important to pay attention to a problem that often emerges in practice. Say we eliminate one of the variables from the model. If one uses listwise deletion to eliminate observations with missing values, default in many software (although not usually SEM software), we may end up with a different sample size between the two models we wish to compare. Again, unless the sample, and hence the sample size, is exactly the same across the two models, they will not be comparable with a nested model test (or even any of the other comparative fit indices). These issues haunt model comparisons for both structural equation and multilevel models and consequently extend also to multilevel structural equation models.

---

[3] The more parsimonious model always fits worse than the more complex one. Therefore, we have to evaluate whether or not it is significantly worse.

[4] This should not be confused with the nested *data* structure of multilevel models discussed later. To differentiate, we use the term *hierarchical* data for what multilevel modelers also call nested data and make use of *nested* only when referring to models. The reader should be aware, though, that in the literature, *nested* is also frequently used for the data structure.

The other two comparative indices, AIC and BIC, are based on the deviance but also include penalties for both model complexity and sample size (see Kline, 2015, ch. 12). The AIC can be expressed as a function of the likelihood (or deviance) and the number of parameters estimated by the model, k (Akaike, 1973), as presented in Equation 1.2:

$$AIC = -2LL + 2k \qquad (1.2)$$

The BIC has similar properties to the AIC but includes an additional penalty for sample size (N) while still taking into account the number of parameters estimated. It can be reformulated as in Equation 1.3:

$$BIC = -2LL + k \ln(N) \qquad (1.3)$$

As with the case of the deviance, both indicators are measures of badness of fit: Lower values denote better-fitting models. Because their absolute numerical values mean little on their own, they are used in a relative fashion: contrasting two nested models with each other to determine which one exhibits a better fit to the data. The same restrictions as described for the likelihood ratio test also stand for comparing AIC and BIC. According to some authors (e.g., Burnham & Anderson, 2002), they can be used *under certain limitations* to compare nonnested models as well.

Unlike multilevel modeling, SEM has been one of the few commonly used modeling techniques in the social sciences that also takes absolute model fit seriously. A whole class of absolute fit indices has been developed to assess how close a model gets to describing reality. These include the χ^2 test, the root mean square error of approximation (RMSEA), Bentler's comparative fit index (CFI), and the standardized root mean square residual (SRMR). These fit measures, the last three of which are directly based on the model χ^2, are unfortunately not applicable to many models in the MSEM framework. The introduction of varying slopes in an MSEM means there is no longer a unique variance-covariance matrix of the model on the basis of which to compute them. For path models that only include random intercepts (as seen in Chapter 2) and most confirmatory factor models where factor loadings do not vary across Level 2 units (as seen in Chapter 3), absolute fit measures are still available. For this reason, we briefly review these fit measures, familiar to those coming from a structural equation modeling background.

The first and most common statistic is the model chi-square (χ^2). A value of zero indicates a model that perfectly fits the data and is obtained with just-identified models. The higher the value, the worse the fit. To know whether the misfit is acceptable, a χ^2 significance test is performed to evaluate whether the estimated variance-covariance matrix is significantly different from the sample observed variance-covariance matrix. If the test is significant, this might indicate a poorly fitting model. However, this test is not definitive, and further ones should always be performed and reported. Among two limitations, it is claimed to be sensitive to larger sample sizes, as well as to fewer degrees of freedom—adding more parameters may reduce the chances of a significant χ^2 test even if the model is not actually improving (Kline, 2015, pp. 270–271).

The χ^2 fit statistic for SEM is therefore not only relevant for comparing models but also interpreted as an absolute model fit. This absolute χ^2 fit statistic may be thought of as a comparison of the model LL estimated through ML to the LL of a baseline unrestricted model that perfectly reproduces the covariance matrix (Kline, 2015, p. 270). This baseline model allows parameters to be estimated up to the point when the number of degrees of freedom is zero ($df_B = 0$) so that the model-implied covariance matrix reproduces the observed covariance matrix. If this difference is statistically significant, it means the model has a significantly higher $-2LL$ than one that can reproduce the variance-covariance matrix and therefore should be rejected for having a bad fit to the data.

More formally, in Equation 1.4, the χ^2 fit statistic is expressed as the product of the sample size (minus 1) and the model fit function that minimizes the discrepancy between observed and expected covariances.[5] The maximum likelihood fit function, F_{ML}, is computed as $log|\Sigma| - log|S| + tr[S\Sigma^{-1}] - p$, where p is the number of covariates (both endogenous and exogenous), S is the observed covariance matrix, and Σ is the model-implied covariance matrix.[6]

$$\chi^2_M = (N-1)F_{ML} \qquad (1.4)$$

Given that a lower F_{ML} indicates a better-fitting model, the discrepancy between observed and expected covariances is closer to 0, and χ^2_M is technically an absolute badness-of-fit index: The higher it is, the worse the fit between model and data. The index is affected by a few characteristics of the data, particularly the sample size and the extent of departure from normality in the distribution of the variables. In the

---

[5] Some software compute this quantity as NF_{ML}.
[6] The formula can be found in Hayduk (1987, p. 137).

latter instance, though, variations of the χ_M^2 that correct for these problems have been devised: the Satorra-Bentler scaled χ^2 and the Satorra-Bentler adjusted χ^2.

RMSEA (Equation 1.5) is another badness-of-fit index commonly used in SEM. Zero indicates best fit, with higher positive numbers denoting progressively worse-fitting specifications. The formula uses a quantity called the "limit of close fit," which is denoted here by $\hat{\delta}_M$. This is based on the fact that a correctly specified model should have $\chi_M^2 = df_M$, irrespective of the sample size. Larger deviations from this equality, $\chi_M^2 > df_M$, will result in higher values of $\hat{\delta}_M$ and thus higher values of the RMSEA index. An often-cited rule of thumb suggests that an $\hat{\varepsilon}$ smaller than 0.05 is indicative of a well-fitting model, while values higher than 0.1 indicate a serious misfit. For single-level models, the RMSEA is estimated with a 90% confidence interval, but that is not the case for multilevel ones. Both the thresholds mentioned earlier (0.05 and 0.1) should be used with a healthy dose of skepticism and always in conjunction with other fit indices.

$$\hat{\varepsilon} = \sqrt{\frac{\hat{\delta}_M}{(N-1)df_M}}, \quad \text{where } \hat{\delta}_M = max(0, \chi_M^2 - df_M) \quad (1.5)$$

The Bentler CFI (Equation 1.6) is a goodness-of-fit index, with values ranging from 0 to 1, where 1 denotes the best fit. It compares the deviation from close fit of our preferred model to the same deviation for a null (empty) model. The null model is one in which covariances between endogenous variables are fixed to 0. Since this is virtually never the case, the null model can be thought of as the worst possible model (Miles & Shevlin, 2007).[7] In the hypothetical instance where our model fits the data perfectly $\chi_M^2 = df_M$, which means $\hat{\delta}_M = 0$, CFI takes on the value of 1. Usually, values of the CFI above 0.95 suggest a good model fit—it indicates that the model is 0.95, or 95% better than the null model.

$$CFI = 1 - \frac{\hat{\delta}_M}{\hat{\delta}_B} \quad (1.6)$$

---

[7] This null model should not be confused with the baseline model in the χ^2 test. There, the baseline model is one in which all covariances between variables are freely estimated. It produces a perfectly fitting model. Here, the null model is one in which all covariances between endogenous variables are fixed to 0. This will usually produce a very poorly fitting model.

The fourth model fit index, the SRMR, is again a badness-of-fit indicator expressed as a function of the average squared covariance residuals in the model, which are the differences between observed and expected covariances. As a result, an SRMR value of 0 is indicative of a perfect-fitting model, with higher values (above 0.07) denoting poor fit. It is important to treat these χ^2 derivative fit values and their associated thresholds for well-fitting models as individual pieces of evidence toward the question of whether our model is appropriate. Each, in isolation, is limited in its ability to conclusively show that a model fits the data well or not. Taken in combination, however, these fit statistics make a stronger case for deciding one way or another with regard to our model, particularly if they point in a similar direction. Some consider that the χ^2 test is the only one that provides strong evidence for or against the appropriateness of a model, since all others are based on it (Barrett, 2007; Hayduk, Cummings, Boadu, Pazderka-Robinson, & Boulianne, 2007; Hayduk & Littvay, 2012). While it is difficult to argue with this position from a statistical point of view, the burden of building fitting models based on this one criterion is extremely high since the χ^2 is sensitive to minor deviations from normality and to increases in sample size. Considering the widespread disregard for model fit in most modeling approaches used in the social sciences, SEM, with its many model fit statistics to consider, is still one of the most conservative model-building approaches.

Further Readings on SEM

The preceding pages briefly introduced the basics of SEM. For more detailed accounts of these topics, including model specification, identification, and estimation, the reader could refer to some of the well-established textbooks. Kline (2015) has an excellent exposition of these topics, as well as further modeling possibilities, with a style accessible for beginners.[8] Bollen (1989) is the foundational work that goes deeper into the mathematics behind SEM but might be more difficult for those not comfortable with matrix algebra.

Readers looking for more advanced applications are referred to the edited volume by Hancock and Mueller (2006), as well as the latter chapters in Hoyle (2012). For a treatment of these models in biology, see Pugesek, Tomer, and Von Eye (2003). To keep in touch with the most

---

[8] We also expect a primer on structural equation models to appear soon, as part of the QASS series.

recent developments in the field, we recommend *Structural Equation Modeling: A Multidisciplinary Journal*, published by Routledge. In the next section, we briefly introduce the "multilevel" building block of MSEM.

Quick Review of Multilevel Models

Multilevel (or MLM, also known as random coefficients or mixed-effects) specifications are classes of regressions that can accommodate independent variables at multiple levels of a data hierarchy in the same model. If we think of a data configuration as comprising multiple levels of analysis (children in schools, regions in European Union countries, voters in electoral districts), then multilevel models allow the simultaneous estimation of the effect of both Level 1 (children, regions, voters) and Level 2 (school, countries, districts) independent variables on the dependent one. Alternatively, we can also view multilevel models as a combination of two specifications: a model for the outcome at the Level 1 and a second model for the *coefficients* from the Level 1, this latter one making use of independent variables at the second level (Gelman & Hill, 2007, p. 1). In political science, multilevel models (henceforth, MLMs) have recently found widespread use, primarily in situations where the interest lies not only in the average effect but also on how a particular effect (say, of education on a person's income) systematically varies between contexts.

We introduce in Figure 1.4 a few more notation conventions for our graphical presentation of MLM models. These conventions extend those found in Figure 1.1 to the domain of multilevel models. We follow L. K. Muthén and Muthén (1998–2017) in using a solid circle to denote a random intercept or slope in our multilevel setup, depending on whether it is placed at the tip or the middle of the arrow, respectively. Finally, a

Figure 1.4 MLM Graphical Notation

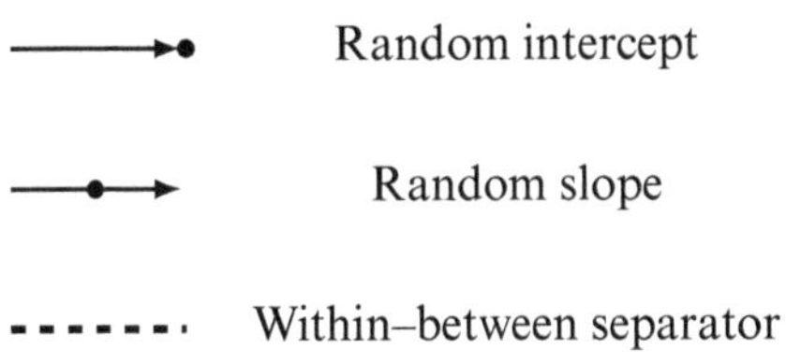

——→•   Random intercept

——•→   Random slope

- - - - - - ·   Within–between separator

thick dashed line establishes the separation between the within-level and between-level part of our multilevel model.[9]

Notation

We can use the standard ordinary least squares (or OLS) regression model as a helpful stepping stone to understanding how MLMs work.

$$Y_i = \beta_0 + \beta_1 X_i + \varepsilon_i, \quad \varepsilon_i \sim \mathcal{N}(0, \sigma^2) \qquad (1.7)$$

In Equation 1.7, we are simply explaining our endogenous variable Y using X and assume that the residuals ε from this model follow a normal distribution with a mean of 0 and constant variance σ^2. As the workhorse model of inferential statistics, OLS regression has undeniable benefits: It is computationally fast, produces unbiased and consistent parameter estimates as well as efficient estimates of sampling uncertainty, and is robust to small violations in assumptions. However, the OLS "machinery" breaks down in situations where the data are clustered into higher-order units, which themselves exert an influence over the phenomenon being studied. In this case, treating the observations as independent leads to a sample size that is deceptively large. In this instance, the effective sample size, or the actual pieces of independent information we should use to calculate standard errors, will be smaller than our total sample size (Snijders & Bosker, 1999, p. 16). The more homogeneous the units, the larger this *design effect* will be and the lower the *effective* sample size compared to the actual sample size (Snijders & Bosker, 1999, p. 23).

We encounter this situation, for example, when we work with data comprising students from multiple schools, and our interest is in studying the covariates of students' test scores. While student-related factors play a major role, so do the practices and conditions at the school level: budget, number of professors with a doctorate, and so on. Let's assume, then, we have sampled j schools from the population of schools in a country, and from this sample of schools, we have once more sampled i students. For these students, we have information on their mathematics test scores (denoted by the label Y). Our MLM model then turns into what we see in Equation 1.8.

$$Y_{ij} = \beta_{0j} + \beta_{1j} X_{ij} + \varepsilon_{ij}, \quad \varepsilon_{ij} \sim \mathcal{N}(0, \sigma^2) \qquad (1.8)$$

---

[9] Throughout the book, we only rely on examples with two levels of analysis, which requires the use of only one line of separation.

We now have an intercept as well as a slope for the effect of X for each of our j schools in the sample. This is a simple extension of the OLS specification, involving just one additional subscript. At the same time, the implications it carries are substantial. We no longer have the same relationship between X and Y across all contexts; rather, this changes from school to school, presumably based on a school-level factor. Collectively, these parameters are assumed to have a normal distribution, with a variance that can be explained by a Level 2 model, like below.

$$\begin{cases} \beta_{0j} = \gamma_{00} + \gamma_{01}Z_j + \upsilon_{0j}, & \upsilon_{0j} \sim \mathcal{N}(0, \sigma^2) \\ \beta_{1j} = \gamma_{10} + \gamma_{11}Z_j + \upsilon_{1j}, & \upsilon_{1j} \sim \mathcal{N}(0, \sigma^2) \end{cases} \quad (1.9)$$

The notation used in Equation 1.9 is borrowed from Raudenbush and Bryk (2002) and makes evident how a second set of γ coefficients, called *hyperparameters* by Gelman and Hill (2007), constitutes statistical models for the Level 1 coefficients. In our particular setting, Z_j represents a school-level independent variable, while the υs are residuals at the school level, which follow the same distributional assumptions as errors from any OLS regression. Plugging in Equation 1.9 into Equation 1.8 gives us the extended form of this multilevel model.

$$Y_{ij} = \gamma_{00} + \gamma_{01}Z_j + \gamma_{10}X_{ij} + \gamma_{11}Z_jX_{ij} + \upsilon_{0j} + \upsilon_{1j}X_{ij} + \varepsilon_{ij} \quad (1.10)$$

The specific model presented here can be encountered in the literature as a random intercepts (β_{0j}) random slopes (β_{1j}) model, which also includes a cross-level interaction (Z_jX_{ij}). The model comprises parameters that are *fixed* (take the same value across the schools) and that are *random* (vary depending on the school). When a multilevel modeler refers to *fixed effects* and *random effects*, they refer to these estimates. In the latter category, we have the υs and the ε, clearly identifiable due to the continued presence of a j subscript even in the extended form of the model. It is this mix of types of parameters that led to the parallel name of "mixed-effects models" (Pinheiro & Bates, 2000). The specification presented in Equation 1.10 represents a barebones version of a random intercepts and random slopes MLM. The structure can, of course, be simplified by restricting the slope to be the same across schools. On the other hand, it can also be complicated further, if required by the theoretical framework we operate with, by adding an additional level of analysis (e.g., school districts or countries), or multiple two-way and even three-way interactions.

Estimation and Model Fit

In the frequentist framework, multilevel models are estimated through maximum likelihood, in either its *full-information* or *restricted* variant.[10] These ML-based estimates carry all the beneficial properties described already in the SEM section. At the same time, Bayesian estimation is gaining popularity, partly due to its flexibility and robustness in small-sample situations (Gelman & Hill, 2007; Stegmueller, 2013). Model building generally proceeds in an incremental fashion. The researcher would start from small models and gradually increase the complexity of the specification by adding exogenous variables, starting with the lowest level and even (*cross-level*) interactions between such variables at different levels. Improvements in model fit would be monitored at each one of the steps using the likelihood ratio test (as presented already in the SEM section).

Sample Size

Discussions of sample size are complicated here by the existence of multiple levels at which data are aggregated, by the diversity of quantities of interest (fixed vs. random effects and their corresponding variance estimates), and by the type of model estimated. A common conclusion, though, is that if the interest is in the fixed effects and their variance, the number of groups is slightly more important than the average group size (Maas & Hox, 2005, p. 88; Raudenbush & Liu, 2000, p. 204). Rules of thumb abound. Kreft (1996) advocates a 30/30 guideline: at least 30 groups with 30 observations in each. On the other hand, if the specification includes cross-level interactions, a more desirable rule would be 50/20, which could even increase to 100/10 if variance components are of interest (Hox, 2010, p. 235). In this latter case, power to estimate random effects is gained from increasing Level 2 sample size to 50 or 100. Most of the rules encountered in the literature cover the case of a continuous outcome. For generalized linear mixed-effects models, though, sample size considerations become more stringent—Moineddin, Matheson, and Glazier (2007) suggest a 50/50 guideline for these types of specifications. The multitude of considerations makes it difficult to distill all the recommendations into universal advice, except to say that most applications

---

[10] While generalized least squares (GLS) estimation was also common in the past, it has since fallen out of favor due to the tendency to produce biased parameter estimates, as well as inefficient standard errors (Hox, 2010, pp. 42–43).

will require a minimum Level 1 sample size of about 10 observations per group (Snijders & Bosker, 1993).[11]
Speaking specifically about Level 2 sample size, though, for a continuous outcome and a specification where the focus is on fixed-effects estimates and standard errors, the minimum indeed appears to be around 30. At the same time, if the interest is only in the Level 2 fixed-effects point estimates, even a sample of 15 clusters might be sufficient for unbiased parameters. On the other hand, if the interest is in a dichotomous outcome, as well as the fixed-effects standard errors, the minimum number of clusters should be around 50 (see McNeish & Stapleton, 2016). Differences exist between estimators themselves, with maximum likelihood underperforming compared to restricted maximum likelihood, particularly with respect to estimates of variance components. Bayesian estimation has been shown to be more robust to small sample sizes at the group level (Stegmueller, 2013) but at the cost of drastically higher computational costs.[12]

Further Readings on MLM

So far, the discussion has revolved only around hierarchical linear models, where a clear and mutually exclusive hierarchical structure could be established: A voter (or student) can come from one and only one country (or school). We have also mainly discussed them in the context of a continuous outcome variable. Readers interested in brushing up on these topics, covered in greater depth than we could do here, are invited to consult the first nine chapters of Raudenbush and Bryk (2002), Snijders and Bosker (2012), Kreft and de Leeuw (1998), or Luke (2004).
The logic of these models, along with the notation, can be extended with added effort to other types of data structures, where individuals can be members of cross-cutting Level 2 units or members in multiple types of units at the same time. Such types of models, called *cross-classified* and *multiple-membership* models, can further be paired with a diverse

---

[11] Common exceptions to this are longitudinal studies, tracking many observations over a few measurement occasions, or household surveys. As long as the Level 2 sample size is sufficient, these designs will allow for precise estimation of fixed effects and their variance; they do, however, affect the ability to estimate random effects (Snijders, 2005).

[12] Recent simulation-based work done by Martin Elff and coauthors even suggests that for small Level 2 sample sizes, restricted maximum likelihood (REML) with an appropriate correction for degrees of freedom (e.g., the Satterthwaite approximation) might be perfectly accurate, obviating the need for Bayesian estimation (Elff, Heisig, Schaeffer, & Shikano, 2016).

array of outcome variables: dichotomous, ordered categorical, multinomial, and counts and survival times. A concise theoretical presentation of these models is given in Raudenbush and Bryk (2002, ch. 10), and a more extensive one can be found in Hox (2010) or Goldstein (2011). For readers interested in the Bayesian approach to MLMs, Gelman and Hill (2007) provide an accessible introduction, as do Gill (2015, ch. 12) and Kruschke (2014, ch. 9). A good presentation of the "frontiers" of MLM can be found in the edited volume of Hox and Roberts (2011).

In the following section, we connect the SEM and MLM modeling approaches and explain the notation used for such models.

Introduction to MSEM and Its Notation

A Multilevel Path Model

Structural equation models are usually presented either in matrix algebra notation or in path diagrams. Some separate out the matrix algebra into multiple simultaneous regression-like equations. For simplicity, in this book, we use the same approach of simultaneous equations and path models. In standard structural equation model form, these simultaneous equations are like simple regressions. In a multilevel structural equation model, these regressions are, in themselves, multilevel models and need to be written as such. We have tried, as much as possible, to maintain the notation used in standard multilevel texts (Raudenbush & Bryk, 2002). Throughout the text, we use β and γ to denote regression coefficients at the Level 1 and Level 2, respectively. Similarly, ε and υ are used to denote residuals at the Level 1 and Level 2, respectively. We follow SEM traditions in using different Greek letters for factor loadings.

In this most general form, Ys indicate endogenous variables. In other words, these are dependent variables with arrows going into them. Xs denote exogenous variables that are only covariates of endogenous variables. These variables are subscripted with i and j denoting that the observation is for the ith person (or other Level 1 unit) in the jth hierarchical unit (whatever the Level 2 units of observations are; e.g., countries).

However, we depart from established multilevel regression notation in our use of superscripts. In all equations, superscripts indicate the endogenous variable that registers the effect being estimated. In a simple multilevel model, we have only one outcome variable. With simultaneous multilevel regression equations, we have multiple multilevel equations with different dependent variables. This is where the superscript clarifies. Using this system, we can write the equation for a simple path model with two Level 1 variables ($X1$ and $Y2$) as in Equation 1.11.

$$\begin{cases} Y1_{ij} = \overset{Y1}{\beta_{0j}} + \overset{Y1}{\beta_{1j}} X1_{ij} + \overset{Y1}{\beta_{2j}} Y2_{ij} + \overset{Y1}{\varepsilon_{ij}} \\ Y2_{ij} = \overset{Y2}{\beta_{0j}} + \overset{Y2}{\beta_{1j}} X1_{ij} + \overset{Y2}{\varepsilon_{ij}} \end{cases} \quad (1.11)$$

$X1$ and $Y2$ are exogenous covariates for $Y1$, while $X1$ is exogenous for $Y2$ itself. With $\overset{Y1}{\beta_{1j}}$ in the first line of Equation 1.11, our notation makes it clear that this is the effect of $X1$ on $Y1$, clearly differentiating it from the β_{1j} in line 2 of Equation 1.11. Using the same syntax rules, a simple Level 2 model for the intercepts at the Level 1 could be the one shown in Equation 1.12.

$$\begin{cases} \overset{Y1}{\beta_{0j}} = \overset{Y1}{\gamma_{00}} + \overset{Y1}{\upsilon_{0j}} \\ \overset{Y1}{\beta_{1j}} = \overset{Y1}{\gamma_{10}} \\ \overset{Y1}{\beta_{2j}} = \overset{Y1}{\gamma_{20}} \\ \overset{Y2}{\beta_{0j}} = \overset{Y2}{\gamma_{00}} + \overset{Y2}{\upsilon_{0j}} \\ \overset{Y2}{\beta_{1j}} = \overset{Y2}{\gamma_{10}} \end{cases} \quad (1.12)$$

Notice how, in Equation 1.11, only the intercepts $\overset{Y1}{\beta_{0j}}$ and $\overset{Y2}{\beta_{0j}}$ are allowed to vary across the Level 2 units. The variances of these intercepts, respectively, are $\overset{Y1}{\upsilon_{0j}^2}$ and $\overset{Y2}{\upsilon_{0j}^2}$. This model does not allow for the variance of the slopes.

Plugging in Equation 1.12 into Equation 1.11 will give us the extended form of the multilevel path model. Due to the presence of multiple equations at the Level 1, this extended model also comprises two equations.

$$\begin{cases} Y1_{ij} = \overset{Y1}{\gamma_{00}} + \overset{Y1}{\gamma_{10}} X1_{ij} + \overset{Y1}{\gamma_{20}} Y2_{ij} + \overset{Y1}{\upsilon_{0j}} + \overset{Y1}{\varepsilon_{ij}} \\ Y2_{ij} = \overset{Y2}{\gamma_{00}} + \overset{Y2}{\gamma_{10}} X1_{ij} + \overset{Y2}{\upsilon_{0j}} + \overset{Y2}{\varepsilon_{ij}} \end{cases} \quad (1.13)$$

A convenient graphical depiction of this random intercepts model is presented in Figure 1.5.[13] Each model presents two sets of figures, one for each level of analysis. Level 1 looks just like any other single-level structural equation model. One augmentation of the standard notation is that parameters that vary between Level 2 units are denoted by solid black dots. These parameters can be intercepts, at the arrowheads (as in Figure 1.5), or they can be slopes, presented in the middle of the arrows: ─•→ (Figure 1.7 below shows a use of our random slopes notation).

On the between level, any parameter on the first, within, level that varies between levels is denoted as a latent variable. It is latent as it technically is not observed at that level of analysis directly; rather, it is estimated from the lower-level information. Of course, as is the case with multilevel regression models, exogenous covariates and, in the case of MSEM, endogenous (outcome) variables can also be entered at this level. Observed variables at this level are still denoted by squares.

A simple extension of this notation generalizes this model to a setting where both intercepts and slopes at the Level 1 are explained by a Level 2

Figure 1.5 Multilevel Path Model Translated to Graphical Form

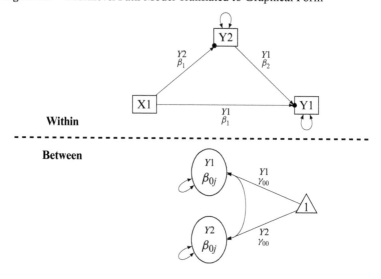

---

[13] Throughout the book, the figure formulation is largely based on L. K. Muthén and Muthén (1998–2017), although the *Mplus* manual does not include a mean structure in any of the graphical models.

variable we call Z. In the interest of expediency, we show the Level 1 and Level 2 models (Equations 1.14 and 1.15), followed immediately by the extended form of the model (Equation 1.16). This extended form is obtained after a few simple multiplication operations, followed by reorganizing of the equation terms. A graphical presentation of the model can be seen in Figure 1.6, which affords the reader the benefit of linking paths with the actual coefficients ultimately estimated in the model.

$$\begin{cases} Y1_{ij} = \beta_{0j}^{Y1} + \beta_{1j}^{Y1} X1_{ij} + \beta_{2j}^{Y1} Y2_{ij} + \varepsilon_{ij}^{Y1} \\ Y2_{ij} = \beta_{0j}^{Y2} + \beta_{1j}^{Y2} X1_{ij} + \varepsilon_{ij}^{Y2} \end{cases} \quad (1.14)$$

Figure 1.6   Multilevel Path Model With Random Intercepts and Slopes Translated to Graphical Form

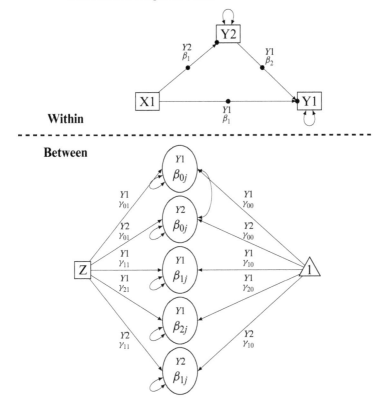

$$\begin{cases} \beta_{0j}^{Y1} = \gamma_{00}^{Y1} + \gamma_{01}^{Y1} Z_j + \upsilon_{0j}^{Y1} \\ \beta_{1j}^{Y1} = \gamma_{10}^{Y1} + \gamma_{11}^{Y1} Z_j + \upsilon_{1j}^{Y1} \\ \beta_{2j}^{Y1} = \gamma_{20}^{Y1} + \gamma_{21}^{Y1} Z_j + \upsilon_{2j}^{Y1} \\ \beta_{0j}^{Y2} = \gamma_{00}^{Y2} + \gamma_{01}^{Y2} Z_j + \upsilon_{0j}^{Y2} \\ \beta_{1j}^{Y2} = \gamma_{10}^{Y2} + \gamma_{11}^{Y2} Z_j + \upsilon_{1j}^{Y2} \end{cases} \quad (1.15)$$

$$\begin{cases} Y1_{ij} = \gamma_{00}^{Y1} + \gamma_{10}^{Y1} X1_{ij} + \gamma_{20}^{Y1} Y2_{ij} + \gamma_{01}^{Y1} Z_j + \gamma_{11}^{Y1} Z_j Y2_{ij} + \\ \quad + \gamma_{21}^{Y1} Z_j X1_{ij} + \upsilon_{1j}^{Y1} X1_{ij} + \upsilon_{2j}^{Y1} Y2_{ij} + \upsilon_{0j}^{Y1} + \varepsilon_{ij}^{Y1} \\ Y2_{ij} = \gamma_{00}^{Y2} + \gamma_{10}^{Y2} X1_{ij} + \gamma_{01}^{Y2} Z_j + \gamma_{11}^{Y2} Z_j X_{ij} + \upsilon_{1j}^{Y2} X_{ij} + \\ \quad + \upsilon_{0j}^{Y2} + \varepsilon_{ij}^{Y2} \end{cases} \quad (1.16)$$

In the following section, we build a new model and introduce the use of latent variables in the multilevel setting.

Full Structural Models in a Multilevel Setting

Full structural models can get quite abstract at first glance. Let us imagine that Figure 1.7 depicts a scenario where students are clustered in classrooms. This takes place in the lower grades of primary school, when pupils often take all of their classes from one teacher. At Level 1, the factors could be two different ability measures, such as math and verbal skills ($FW1$ and $FW2$), each measured with three different tests ($Y1$–$Y6$). The factors may correlate but attest to distinct students' abilities. They both can be predicted by an observed student characteristic (XW), such as the number of hours doing homework, and they can also predict an observed student-level variable (YW), like interest in school activities.

At the between level, FB is classroom teachers' ability to teach. Here, there is less theoretical reason to separate various subjects, especially in lower grades, so a single factor can convey that ability. This single factor is indicated by the class averages of student ability measures in each of the six tests. At this level of analysis, there could be a teacher-level outcome and exogenous variable (YB and ZB), not only associated with the teacher-level factor but also the variance in slopes

Figure 1.7 Full SEM Model With Random Intercepts and Random Slopes Translated to Graphical Form

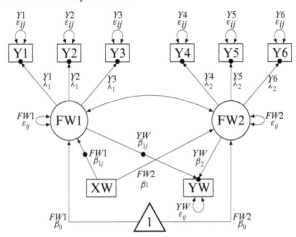

Within

Between

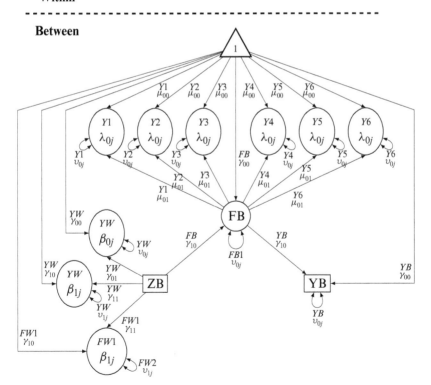

of within-level relationships, and variation of within-level intercept(s) across classrooms.

In Figure 1.7, within factors are measured by three indicators each ($Y1$–$Y6$). This part is described in Equation 1.17. Ys are the observed indicators, which can be decomposed into an intercept λ_{0j} that varies across groups j, factor loadings λ_1 or λ_2, invariant across groups and individuals, that multiply the latent constructs $FW1$ or $FW2$, which vary across individuals i and groups j. Finally, there is an error term ε_{ij} associated with each indicator, varying for each individual. The numerical subscript in λ is the latent variable to which it refers. λ_1 are factor loadings of $FW1$, while λ_2 are factor loadings of $FW2$.[14]

$$\begin{cases} Y1_{ij} = \lambda_{0j}^{Y1} + \lambda_1^{Y1} FW1_{ij} + \varepsilon_{ij}^{Y1} \\ Y2_{ij} = \lambda_{0j}^{Y2} + \lambda_1^{Y2} FW1_{ij} + \varepsilon_{ij}^{Y2} \\ Y3_{ij} = \lambda_{0j}^{Y3} + \lambda_1^{Y3} FW1_{ij} + \varepsilon_{ij}^{Y3} \\ Y4_{ij} = \lambda_{0j}^{Y4} + \lambda_2^{Y4} FW2_{ij} + \varepsilon_{ij}^{Y4} \\ Y5_{ij} = \lambda_{0j}^{Y5} + \lambda_2^{Y5} FW2_{ij} + \varepsilon_{ij}^{Y5} \\ Y6_{ij} = \lambda_{0j}^{Y6} + \lambda_2^{Y6} FW2_{ij} + \varepsilon_{ij}^{Y6} \end{cases} \quad (1.17)$$

The higher level of the measurement model is seen in Equation 1.18. The between-level factor is measured by all six Y indicators. In MSEM, the group means of indicators from the within level are used to estimate factor loadings at the between level, across groups. For this reason, $Y1$ through $Y6$ are denoted in the between part of Figure 1.7 by λ_{0j}^{Y1}–λ_{0j}^{Y6}. Furthermore, they are represented by circles since they are parameters estimated in the within part of the model, not directly observed. At the within part, they are indicated by the solid circles at the end of arrows parting from $FW1$ and $FW2$ to $Y1$ through $Y6$.

In Equation 1.18, each indicator intercept from Level 1 (λ_0) is estimated by the overall intercept across Level 2 units (μ_{00}), factor loadings μ_{01} that multiply FB and the variance of intercepts v_{0j}^{Y1-6} across groups.

---

[14] Both factor loadings and factor means can also have between-group variance. That notation follows what is described here, and such models are discussed and presented in Chapters 3 and 4.

For this reason, the triangle denoting the intercept at Level 1 has arrows only to the two endogenous variables that have no Level 2 variance: $FW1$ and $FW2$. The other intercepts are all denoted as varying across groups by black dots. At the between part, all of these overall intercepts are then the arrows from the triangle.

$$\begin{cases} \lambda_{0j}^{Y1} = \mu_{00}^{Y1} + \mu_{01}^{Y1} FB_j + \upsilon_{0j}^{Y1} \\[4pt] \lambda_{0j}^{Y2} = \mu_{00}^{Y2} + \mu_{01}^{Y2} FB_j + \upsilon_{0j}^{Y2} \\[4pt] \lambda_{0j}^{Y3} = \mu_{00}^{Y3} + \mu_{01}^{Y3} FB_j + \upsilon_{0j}^{Y3} \\[4pt] \lambda_{0j}^{Y4} = \mu_{00}^{Y4} + \mu_{01}^{Y4} FB_j + \upsilon_{0j}^{Y4} \\[4pt] \lambda_{0j}^{Y5} = \mu_{00}^{Y5} + \mu_{01}^{Y5} FB_j + \upsilon_{0j}^{Y5} \\[4pt] \lambda_{0j}^{Y6} = \mu_{00}^{Y6} + \mu_{01}^{Y6} FB_j + \upsilon_{0j}^{Y6} \end{cases} \quad (1.18)$$

The structural component of the model includes one within-level exogenous covariate for the latent variables, XW, and one within-level endogenous variable YW, explained by the latent variables $FW1$ and $FW2$. At the higher level, there is an exogenous covariate ZB and an outcome YB. In addition, we estimate a random intercept of YW and random slopes for the effects of $FW1$ on YW, and of XW on $FW1$, as indicated by the small solid circles in the within part of Figure 1.7 and residuals for all endogenous variables. The within-level structural part of the model is given in Equation 1.19.

$$\begin{cases} YW_{ij} = \beta_{0j}^{YW} + \beta_{1j}^{YW} FW1_{ij} + \beta_{2}^{YW} FW2_{ij} + \varepsilon_{ij}^{YW} \\[4pt] FW1_{ij} = \beta_{0j}^{FW1} + \beta_{1j}^{FW1} XW_{ij} + \varepsilon_{ij}^{FW1} \\[4pt] FW2_{ij} = \beta_{0j}^{FW2} + \beta_{1}^{FW2} XW_{ij} + \varepsilon_{ij}^{FW2} \end{cases} \quad (1.19)$$

The corresponding between-level structural model is presented in Equation 1.20.

$$\begin{cases} YB_j = \gamma_{00}^{YB} + \gamma_{10}^{YB} FB_j + \upsilon_{0j}^{YB} \\ FB_j = \gamma_{00}^{FB} + \gamma_{10}^{FB} ZB_j + \upsilon_{0j}^{FB} \\ \beta_{0j}^{YW} = \gamma_{00}^{YW} + \gamma_{01}^{YW} ZB_j + \upsilon_{0j}^{YW} \\ \beta_{1j}^{YW} = \gamma_{10}^{YW} + \gamma_{11}^{YW} ZB_j + \upsilon_{1j}^{YW} \\ \beta_{1j}^{FW1} = \gamma_{10}^{FW1} + \gamma_{11}^{FW1} ZB_j + \upsilon_{1j}^{FW1} \end{cases} \quad (1.20)$$

For completeness, we make clear in Equation 1.21 that two slopes and two intercepts are not allowed to vary.

$$\begin{cases} \beta_{2j}^{YW} = \gamma_{20}^{YW} \\ \beta_{0j}^{FW1} = \gamma_{00}^{FW1} \\ \beta_{0j}^{FW2} = \gamma_{00}^{FW2} \\ \beta_{1j}^{FW2} = \gamma_{10}^{FW2} \end{cases} \quad (1.21)$$

The notation we have chosen is admittedly sui generis, but we believe it maximizes ease of understanding. Alternative systems posed considerable difficulties. Relying entirely on existing multilevel notation would have led to confusion generated by the possibility of multiple Level 1 equations in an MSEM framework, each with its own β_{0j}. SEM approaches, on the other hand, usually resort to matrices as the better way to accommodate several simultaneous equations. For example, using the notation from B. O. Muthén (1994, p. 382), Equations 1.17 and 1.18, defining the measurement model, could be rewritten as shortly as seen in Equation 1.22.

$$Y_{ij} = \nu + \Lambda_B \eta_{Bj} + \varepsilon_{Bj} + \Lambda_W \eta_{Wij} + \varepsilon_{Wij} \quad (1.22)$$

In the formulation above, Y_{ij} is the vector of six indicators, varying for individuals i within j groups. ν is a vector of intercepts; Λ_B and Λ_W are matrices of factor loadings for the between and within parts of the model, respectively; and η_{Bj} and η_{Wij} are the matrices of latent variables in the two parts. ε_{Bj} and ε_{Wij} are the vectors of residuals for the between and within parts. This notation is succinct but more complicated to grasp for newcomers to such models. Our choice highlights every single parameter estimated in the models, which can be matched to software outputs.

Estimation and Model Fit

Given that MSEM models are, at their core, structural equation specifications with data on multiple levels of analysis, the estimation and the measures of fit produced by it are similar to those described in the SEM estimation section. Models should be compared with each other by examining the model χ^2, RMSEA, CFI, AIC, or BIC or simply by conducting a likelihood ratio test on the $-2LL$, while keeping in mind that on occasion, these might point to different conclusions.[15]

Using these indices for the full model, however, might be problematic for a few reasons identified by Yuan and Bentler (2007): First, in case of bad fit, it is not possible to know whether the badness of fit comes from the within or the between part. Second, in many applications, the within-level sample size is much larger than the between-level one. The ML fitting function weights the models differently depending on the sample size in each level, so that in case the Level 1 sample is much larger, the χ^2 test of exact fit and the indices that build on it (RMSEA, CFI, and the Tucker-Lewis Index [TLI]) are dominated by the Level 1 model. Of those fit indices presented earlier, the only one giving separate within and between fit is the SRMR, which can be used to assess misspecification at a specific level of analysis.

Furthermore, there is one extension of the other existing fit indices to MSEM: the partially saturated model test (Ryu & West, 2009). For this test, the researcher first specifies the hypothesized model for the within level, as well as a saturated model for the between part, calculating the χ^2 and other fit indices. Since the between model is saturated and therefore fits perfectly, any badness of fit is respective to the within level. Next the models are switched: The between-level model is specified as hypothesized, and a saturated model is specified at the within level. Conversely, any badness of fit indicated by the χ^2 test, CFI, TLI, and RMSEA is the between-level fit. This approach gives a more precise

---

[15] In MSEM, absolute fit indices (χ^2 test, RMSEA, SRMR, and CFI), however, cannot currently be computed for models with random slopes.

estimation of problems with model fit and can indicate whether one of the levels (or both) is perhaps misspecified. We demonstrate the partially saturated model test in Chapter 3 on only one example of a multilevel measurement model. For the remaining examples in the book, we rely on global fit and the SRMR for indicating possible misfit across the two levels. To date, there is no automated implementation of the partial saturation test, and fitting three separate models for each example would make the demonstration of other topics distractingly tedious. Researchers are advised, however, to assess model fit on both levels of analysis in their own studies.

Scope of the Book and Online Materials

In the next chapter, we demonstrate the most common uses of multilevel path models tapping into a cross-national survey. In Chapter 3, we demonstrate multilevel confirmatory factor models. We do this using an education data set that includes surveys of students clustered within schools. Finally, in Chapter 4, we demonstrate the most common multilevel structural equation models that include both latent variables and structural paths between these latent (and observed) variables, using a data set of employees grouped into companies. Concluding remarks in Chapter 5 include brief discussions of potential MSEM modeling extensions.

This book also has an online appendix, hosted at the following address: http://levente.littvay.hu/msem/. It includes data and replication scripts to reproduce all models presented in Chapters 2, 3, and 4. All models were developed to work with, and all results were obtained using the free demo version of *Mplus* 8.1 (L. K. Muthén & Muthén, 1998–2017), available at http://statmodel.com. For some models, we also provide R scripts that run with the package lavaan, Version 0.6-2 (Rosseel, 2012).[16] As of this writing, lavaan cannot run multilevel structural equation models with random slopes (i.e., Figures 2.5, 3.5, and 4.8). If and when this changes, the appendix will be updated to include those examples. We also provide LATEX code to produce all figures and equations in this book. Any errors that will inevitably be identified will also be mentioned in a document uploaded on the web page.

---

[16] We owe a debt of gratitude to Yves Rosseel for translating our *Mplus* models into R.

CHAPTER 2. MULTILEVEL PATH MODELS

In the previous section, we offered a quick overview of both structural equation models and multilevel models, as well as how both the modeling and the notation can be merged into a single framework. Up to this point, the most common method for the analysis of hierarchical data structures was limited to regression-like modeling situations. This chapter generalizes a more complex structure of relationships to a multilevel framework. Path models are arguably the most simple structural equation models, incorporating only observed variables but going beyond the situation with only one endogenous variable and multiple exogenous ones. Here, we will generalize such models to the multilevel framework.

Simplicity often goes hand in hand with flexibility, and path models have been used for a wide range of questions. They have proved to be especially powerful and insightful when combined with a strong argument for what the proper causal ordering should be among variables in a model. A case in point is Blau and Duncan's (1967) celebrated account of intergenerational mobility within the occupational hierarchy of the United States. The authors make a strong case for parental education and occupation as pure exogenous factors, which come to influence offspring's education and their first job (endogenous factors). Through these transmission mechanisms but also directly, they come to shape the second generations' current position in the occupational hierarchy (see Blau & Duncan, 1967, fig. 5.1). Through the use of path modeling, the authors are able to determine the relative contribution of parental factors and of personal effort to the process of occupational stratification, as well as the pathways through which parental factors operate.

A further feature of this modeling strategy is its ability to capture reciprocal effects via nonrecursive specifications. An early example of this is the Duncan, Haller, and Portes (1971) model of the influence of peers on professional aspirations. Here, both personal intelligence and family socioeconomic status (SES) shape a person's professional aspirations. The corresponding pair of factors naturally shape a peer friend's aspirations. However, the authors also introduce the potential for a friend's family SES to influence aspirations through the influence of role models, as well as for a friendship dyad's aspirations to shape each other. Parsimonious and elegant, this and the following specifications used by the authors allow them to disentangle how much of a person's professional aspirations are due to personal factors, role models' influence, or peer examples. A final example reveals an oft-forgotten strength of path modeling: the ability to estimate off-diagonal cells in a variance-covariance matrix under the assumption of a properly specified model and then use them further on in the estimation of the model. This is done by Duncan (1968) through the use of multiple sources of data with the goal of estimating cells in the variance-covariance matrix. For the cells where no data source could be used to obtain a measurement due to missing data, an application of path analysis rules produces these covariances based on information already available in the matrix and the model specification (Duncan, 1968, p. 7). As a final step, the model is then estimated using the variance-covariance matrix. These features of path analysis are very portable and powerful in situations where reciprocal effects or associations between predictor variables are suspected to operate.[1]

Before getting to the substance of the chapter, we urge the reader to revisit the notation conventions presented in Figure 1.1 for SEM models and in Figure 1.4 for MLM models. These ways of graphically describing an MSEM model will be used consistently from now on in most of the specifications we discuss. We also note here that while we sometimes use the terms *predictor* and *predict* to refer to exogenous covariates of a variable and their effect, we do not imply a causal ordering through the use of this language.

We start with an example from Wave 4 of the *World Values Surveys* (WVS), a cross-cultural survey incorporating a reasonably large number of countries for multilevel analysis. Our interest is in citizens' self-expression values (available in the WVS data), which occupy a central mediating role in a theorized chain of associations that starts with economic development and ultimately produces democratization (see Inglehart & Welzel, 2009). Such values signal that the individual assigns high importance to participation in decision making, to expression of one's individuality as opposed to conformism, to environmental responsibility, and to tolerance of alternative lifestyles. As a theoretical curiosity, but also as a practical question, we are interested in who are the individuals most likely to exhibit high levels of such self-expression values. Given their heightened predisposition to press authoritarian regimes for increasing political openness, identifying individual-level

---

[1] The reader can find more examples like these in Wolfle (2003).

Table 2.1 List of Variables From *World Values Surveys* Example

Code	Item	Response Scale
Individual level		
SEV	Emphasis on the importance of civic activism, subjective well-being, tolerance and trust, personal autonomy, and choice (Inglehart & Baker, 2000)	Scale approx. ranging in the sample from −1 to 3.16
AGE	Respondent's age	Numeric, ranging from 15 to 101
INC	Respondent's household income	Ordinal scale, ranging from 1 to 10 (deciles)
EDU	Respondent's highest educational level	Ordinal scale, ranging from 1 ("inadequately completed elementary") to 8 ("university with degree")
Country level		
GDP	Country's gross domestic product/capita, adjusted by purchasing power parity, in current international dollars	Numeric, ranging in the sample from approx. 1,000 to 75,000

Note: Original variable names from the WVS data set are survself (*SEV*), X003 (*AGE*), X047 (*INC*), and X025 (*EDU*). Original variable name from QoG data set, January 2016 version, is wdi_gdppcpppcur (*GDP*).

factors connected to such values can help us better explain the appearance of pressures for change in a country. At a deeper level of analysis, also identifying the systemic characteristics associated with a greater preponderance of such values could help advocacy organizations better target their democracy promotion efforts.

The multilevel data structure at hand is one where individuals are grouped within countries. The variables at Level 1 are, first, a constructed scale of self-expression values, available in the original data set; here, higher scores denote an individual's greater emphasis on self-expression. We also have in the data an individual's income, ranging from 1 to 10 (in income deciles), as well as age (measured in decades and rescaled, so that 0 denotes 1.8 decades) and the highest educational level attained (a variable with eight ordered categories). At the country level, gross domestic product (GDP) per capita, adjusted by purchasing power parity (PPP), expressed in constant international dollars, is obtained from the *World Development Indicators*. Following conventions in the literature, the natural logarithm of GDP was used to eliminate model convergence issues, achieve normality of the variable and potentially also the residuals, and ensure that the relationships are closer to linear. In the interest of simplicity, all models were estimated on a sample of 42,619 respondents from 55 countries. This was ensured by performing listwise deletion for missing information prior to estimation.

Our example is admittedly pared down, with only four exogenous covariates for self-expression values. The primary reason for this is the need to keep model specifications and figures at a manageable size and complexity. At the same time, though, our specification captures the core features of most real-life MSEMs. As the reader will see in the following sections, even with such a reduced model, we are able to offer tentative answers to what are the individual-level and country-level factors associated with high self-expression values. It is important to note that the example could be expanded with the addition of further constructs of interest into the model, but the amount of results to be interpreted would grow along. In summary, we find the specification to be a suitable teaching device but falls far short of a causally identified and properly specified model for self-expression values at the individual level.

Multilevel Regression Example

When confronted with a hierarchical data structure and with the previously mentioned set of theoretical questions, the first instinct of the applied data analyst would likely be to use a standard multilevel model. With this data configuration, the analyst can be confident that the standard errors produced by the model are accurate and that any effect of contextual exogenous covariates on the individual-level dependent variable (DV) would be accurately estimated. A very simple model, for demonstration purposes, might be to regress self-expression values on income, education, and age at the individual level and GDP per capita at the country level. This is precisely the type of model used by Welzel and Inglehart (2010) in their investigation of what drives self-expression values.

A presentation of the model in diagram form, along with the estimated parameters, can be seen in Figure 2.1. The results are plausible, albeit based on an underspecified model. All three exogenous covariates are statistically significant, with effects in the expected direction:

Figure 2.1 Standard Multilevel Regression Model Specification (unstandardized results reported)

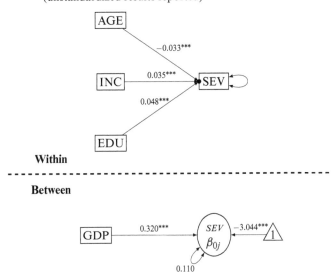

Individuals with a higher level of education exhibit, on average, a greater extent of self-expression values, as do individuals with higher incomes. In a similar way, older individuals display lower levels of such values, although we are unable to say based on this specification whether we are dealing with an age or a cohort effect. More important, although we only use one wave of the WVS, the effects we find are in the range of what Welzel and Inglehart (2010) find, even though their model is slightly more complex and is estimated on three WVS waves. In addition, at the Level 2, GDP per capita *positively* affects the extent to which an individual manifests self-expression values: Wealthier countries also display higher average levels of self-expression values in the population. Further work certainly awaits the applied modeler: testing alternative specifications, arriving at a best-fitting model, inspecting residuals, and so on. Should the final model pass all "quality checks," however, the investigation of the modeler concludes with the final interpretation of coefficients.

It is important to point out that the alternative specifications tried by our researcher all presume direct associations between each exogenous covariate and the outcome, and all estimate direct effects of these variables on the outcome. In many situations, though, this approach clearly ignores linkages that can exist among exogenous variables themselves.

In standard analyses of turnout at the individual level, education is an explanatory factor for turnout and for political efficacy, itself a determinant of turnout. It is also plausible to conceive of social class as directly explaining party choice, due to exposure to party mobilization efforts or pressure from social networks, as well as predicting issue position, which itself comes to guide party choice. Finally, in our simplified example, age is not solely an explanatory factor for self-expression values but also for the level of education attained.

In the case of such direct and indirect statistical associations, the standard modeling approach for the past four decades has been to employ a structural equation model. In such a specification, income, education, and age could not only explain self-expression values but also be connected to each other (e.g., age to education). This effectively turns education from a purely exogenous predictor to an endogenous one. While the theory and estimation routines behind the standard SEM approach are solidly established, these would be of only limited use in our case. The presence of data on multiple level of analysis, with individuals clustered in countries, means that standard errors will be inefficient, and by implication, significance tests produced by the standard SEM toolbox would be imprecise. Using the example of GDP per capita, our analysis would treat all 42,619 members of the sample as contributing unique pieces of information to the final estimated quantity and its uncertainty. However, this is clearly wrong, as for *GDP* we only have 55 measurements, one for each country in our sample. While not as serious, the same problem plagues the estimates for other exogenous and endogenous variables in our model.

The analyst is then faced with a dilemma: either obtain accurate estimates of effect and uncertainty through an MLM model, at the cost of ignoring the larger structure of associations in the model, or model this structure properly through the use of a SEM, at the cost of estimates that ignore the data clustering. In the following sections, we present a few model specifications that allow the researcher to overcome the dilemma posed here and add additional modeling flexibility through the inclusion of variables (both causes and consequences) at the second level of analysis.

Random Intercepts Model

The model specification upon which we base our initial discussion is presented in Figure 2.2, depicting a standard setup for a multilevel path model. A respondent's position on the self-expression values scale is regressed on income, age, and education, while age is a covariate of

Figure 2.2 Standard Path Model

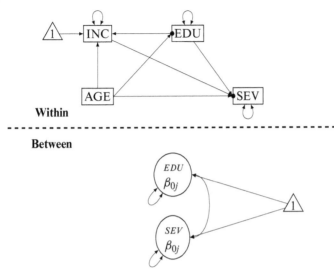

education, and education and age are explanatory factors for a person's income. In this model, therefore, income, education and self-expression values are endogenous variables while age is exogenous. The structure of relationships is presented in notation form in Equation 2.1. In a sense, this is a mediation model (Iacobucci, 2008) where the impact of education and age on self-expression values is mediated by income and education, respectively. In this model, we also control for the impact of age on income, which only has an indirect effect on self-expression.

$$\begin{cases} SEV_{ij} = \overset{SEV}{\beta_{0j}} + \overset{SEV}{\beta_{1j}} INC_{ij} + \overset{SEV}{\beta_{2j}} EDU_{ij} + \overset{SEV}{\beta_{3j}} AGE_{ij} + \overset{SEV}{\varepsilon_{ij}} \\ INC_{ij} = \overset{INC}{\beta_{0j}} + \overset{INC}{\beta_{1j}} EDU_{ij} + \overset{INC}{\beta_{2j}} AGE_{ij} + \overset{INC}{\varepsilon_{ij}} \\ EDU_{ij} = \overset{EDU}{\beta_{0j}} + \overset{EDU}{\beta_{1j}} AGE_{ij} + \overset{EDU}{\varepsilon_{ij}} \end{cases} \quad (2.1)$$

At this point, the model at hand is nothing more than a structural equation model. But we know that individuals in the data set are from their respective countries. To account for the potential bias that emerges from hierarchical data, if ignored, we allow the intercepts in this model to vary across countries. We believe that although the impact of age on education and that of education on self-expression values is roughly the same for each country, the baseline levels of education and self-expression values are different between countries. We hope that most readers would consider this statement plausible, at least with respect to education. For this reason, these intercepts are allowed to vary across countries. It could also be argued that the variances of income could be interesting, but given the standardization of the variable into deciles, the interpretation of the effect (especially on the country level) would be quite difficult. Hence, for the purposes of this exercise, we are treating this variable as a control and not as one with substantive interest, and we are omitting all estimated random effects.

Here we allow the variance of the above mentioned intercepts at the between-country level. By doing so, in MSEM estimation, we split the total covariance matrix into two: one within and one between clusters. They are additive (meaning, the total covariance is the sum of within and between covariances) and uncorrelated (B. O. Muthén, 1994). We also add a predictor on the country level to explain variation in both of the varying intercepts: GDP per capita. The six slopes at the Level 1 (one for income, two for education, and three for age) are not allowed to vary between countries. These relationships are presented in Equation 2.2, where we follow the convention introduced by Snijders and Bosker (1999) of denoting fixed intercepts and slopes with a Level 2 γ rather than a Level 1 β.

$$\begin{cases} \beta_{0j}^{SEV} = \gamma_{00}^{SEV} + \gamma_{01}^{SEV} GDP_j + \upsilon_{0j}^{SEV} \\ \beta_{1j}^{SEV} = \gamma_{10}^{SEV} \\ \beta_{2j}^{SEV} = \gamma_{20}^{SEV} \\ \beta_{3j}^{SEV} = \gamma_{30}^{SEV} \\ \beta_{0j}^{INC} = \gamma_{00}^{INC} \\ \beta_{1j}^{INC} = \gamma_{10}^{INC} \\ \beta_{2j}^{INC} = \gamma_{20}^{INC} \\ \beta_{0j}^{EDU} = \gamma_{00}^{EDU} + \gamma_{01}^{EDU} GDP_j + \upsilon_{0j}^{EDU} \\ \beta_{1j}^{EDU} = \gamma_{10}^{EDU} \end{cases} \quad (2.2)$$

The extended form of the model, for each Level 1 equation, then becomes the specification shown in Equation 2.3.

$$\begin{cases} SEV_{ij} = \gamma_{00}^{SEV} + \gamma_{10}^{SEV} INC_{ij} + \gamma_{20}^{SEV} EDU_{ij} + \gamma_{30}^{SEV} AGE_{ij} + \\ \quad + \gamma_{01}^{SEV} GDP_j + \upsilon_{0j}^{SEV} + \varepsilon_{ij}^{SEV} \\ INC_{ij} = \gamma_{00}^{INC} + \gamma_{10}^{INC} EDU_{ij} + \gamma_{20}^{INC} AGE_{ij} + \varepsilon_{ij}^{INC} \\ EDU_{ij} = \gamma_{00}^{EDU} + \gamma_{10}^{EDU} AGE_{ij} + \gamma_{01}^{EDU} GDP_j + \upsilon_{0j}^{EDU} + \varepsilon_{ij}^{EDU} \end{cases} \quad (2.3)$$

For convenience, the model is also presented in graphical form in Figure 2.3, with estimates from the model included. Much like the simple multilevel model, we see that age exerts a direct effect on self-expression values ($\gamma_{30}^{SEV} = -0.033^{***}$). But it is also clear that some originally independent variables of SEV turn into endogenous variables themselves; for instance, age has a significant impact on education ($\gamma_{10}^{EDU} = -0.297^{***}$).

Figure 2.3 Multilevel Path Model With Random Intercepts (unstandardized results reported)

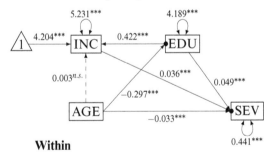

Within

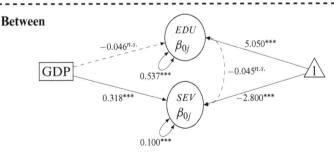

Between

This potentially accounts for some indirect effects the multilevel model completely misses, since it only estimates direct relationships between each covariate and the outcome. Here, this is clearly not the case. The negative effect of age on education is likely explained by the fact that educational opportunities have only recently expanded in a substantial number of countries in our sample. This means that younger people in the population are, on average, more educated than those who are older, leading to the negative estimate we observe.

Moving on to the between level, at the bottom part of Figure 2.3, in this model specification, GDP per capita has a positive impact on the intercept of self-expression values (or, we can also say, directly affects self-expression values) but does not affect educational attainment. It would appear, then, that wealthier countries have a higher level of self-expression values, even after we control for individual-level factors, including income. However, this effect is not due to higher average levels of education in these countries, as richer and poorer countries have roughly similar levels of educational achievement. One explanation might reside in the type of education that richer and poorer countries tend to emphasize. We speculate that wealthier countries emphasize a liberal arts education to a greater degree than do poorer ones. This promotes some of the attitudes that constitute the self-expression cluster, such as self-expression and importance allocated to participation. Lower GDP countries, on the other hand, emphasize this type of education less. In turn, their curriculum is geared more toward memorization and exact sciences. To sum up, we suspect that it is the content of education that is different in countries at different GDP levels, rather than the absolute number of years of education.

For full disclosure, we need to note that the figure has one additional estimated parameter that is not represented in the equation in the interest of parsimony and simplicity. This is the covariance between the intercepts of *EDU* and *SEV* that are allowed to vary across Level 2 units, depicted in the between part of Figure 2.3 by the curved arrow connecting the circles (latent indicators). What is important to remember is that covariances between these Level 2 latent variables that emerge from the variance components of Level 1 intercepts (or slopes, as seen in the next section) can be allowed to covary or they can be fixed to 0, forcing no relationship. We believe the decisions should rest on theoretical expectations, although another school of thought suggests that seeking the appropriate balance between parsimony and model fit should drive these decisions, if need be, atheoretically.

We are confident in the added power of MSEM specifications in the case of hierarchical data structures, compared to MLMs or SEMs, irrespective of the actual conclusions drawn from the data. It is true that the individual-level effects of age, education, and income are roughly similar to those displayed in the previous model. At the same time, such a model specification allows us to offer a richer description of the effect pathways that operate in reality. We now see that age has both a direct effect on self-expression values, as well as an indirect effect, through education. The same can be said of the effect of education on self-expression values.

Random Slopes Model

The model specification presented in Figure 2.3 captures a snapshot of the relationships between self-expression values and age, income, and education, along with GDP per capita at the country level. The key insight from the model is that GDP per capita is associated with the level of self-expression values in a country, suggesting that bottom-up pressures for democratic change could be more likely in wealthier countries (Epstein, Bates, Goldstone, Kristensen, & O'Halloran, 2006; but see Przeworski, Alvarez, Cheibub, & Limongi, 2000 for findings that go against this assertion). Furthermore, within countries, it is younger and more educated citizens who are more likely to harbor these values.

Nevertheless, we contend that there is yet more to discover about the dynamics comprising the data-generating process. Breaking down the relationships between age, education, and self-expression values, we find heterogeneity in the effects of age and education (see Figure 2.4). While

Figure 2.4 Bivariate Relationships Between Education, Self-Expression Values, and Age (individual country fit lines)

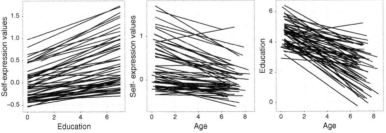

the effect of education on self-expression values is mostly positive, it clearly has varying strength, with a maximum of 0.162 in the Netherlands and a minimum of –0.005 in Uganda (India is the only other country where this effect is negative). In a similar manner, for education, the effect is predominantly negative, with a minimum of –0.162 in Denmark. At the same time, there are clearly situations where this effect is positive, such as Algeria, Hungary, Moldova, or the United States (six countries in total). Finally, the effect of age on education is predominantly negative: Older individuals are, on average, less educated, presumably due to the more restricted educational opportunities available to them when they were transitioning to adulthood. The strongest negative effect is found in Algeria (–0.876). There are, however, cases where this effect is positive (United States), as well as contexts with virtually no effect (the Czech Republic, Tanzania, or Uganda). As a result of this, in our final specification, we have also allowed these relationships to (randomly) vary across countries and have added GDP per capita as an explanatory factor for this variance. GDP per capita is a potential moderator for the slope of education on self-expression values, for the slope of age on self-expression values, and for the slope of age on education. In addition, as in the previous specification, the intercepts of education and self-expression values have been allowed to vary across countries and are predicted by GDP per capita as well. This is actually necessary; when a slope is allowed to vary across the Level 2 units, it is important to allow the intercept to vary as well. For this reason, a random slopes model is also always a random intercept model despite the fact that we just call it a random slopes model for short.

Figure 2.5 presents a graphical depiction of this model, along with the results of the estimation procedure. In Equation 2.4, we only present the extended form of the specification.

$$\begin{aligned}
SEV_{ij} &= \gamma_{00}^{SEV} + \gamma_{10}^{SEV} INC_{ij} + \gamma_{20}^{SEV} EDU_{ij} + \gamma_{30}^{SEV} AGE_{ij} + \\
&+ \gamma_{21}^{SEV} GDP_j EDU_{ij} + \gamma_{31}^{SEV} GDP_j AGE_{ij} + \gamma_{01}^{SEV} GDP_j + \\
&+ v_{2j}^{SEV} EDU_{ij} + v_{3j}^{SEV} AGE_{ij} + v_{0j}^{SEV} + \varepsilon_{ij}^{SEV} \\
INC_{ij} &= \gamma_{00}^{INC} + \gamma_{10}^{INC} EDU_{ij} + \gamma_{20}^{INC} AGE_{ij} + \varepsilon_{ij}^{INC} \\
EDU_{ij} &= \gamma_{00}^{EDU} + \gamma_{10}^{EDU} AGE_{ij} + \gamma_{01}^{EDU} GDP_j + \gamma_{11}^{EDU} GDP_j AGE_{ij} + \\
&+ v_{1j}^{EDU} AGE_{ij} + v_{0j}^{EDU} + \varepsilon_{ij}^{EDU}
\end{aligned}$$

(2.4)

Figure 2.5 Multilevel Path Model With Random Intercepts and Slopes (unstandardized coefficients)

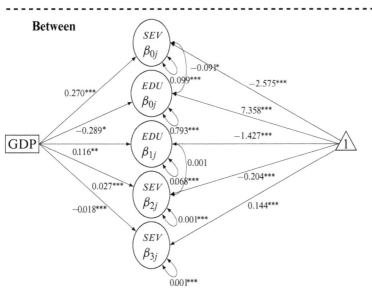

The first change, as compared to the random intercepts model, is that the impact of *GDP* on education becomes significant. Allowing the slopes to vary also affected the variance of the intercepts of education, resulting in an improved ability of *GDP* to explain this variance ($\gamma_{01}^{EDU} = -0.289^*$). GDP per capita also has a statistically significant moderation effect on the impact of education on self-expression values ($\gamma_{21}^{SEV} = 0.027^{***}$), as well as on the slope of age when it is regressed on education ($\gamma_{01}^{EDU} = 0.117^{**}$). Finally, *GDP* also significantly influences the relationship between age and self-expression values ($\gamma_{31}^{SEV} = -0.018^{***}$).

For example, while in extremely poor countries, the effect of education on self-expression values is negative, the effect turns positive in wealthier countries. At the same time, though, in poorer countries, the effect of age on education is negative and relatively small. In wealthier countries, the effect of age on education is larger and even further in a negative direction. In essence, in wealthier countries, a reasonably large amount of education is the norm today, whereas this was not the case a mere five decades ago. For this reason, older people often have substantially less education than do younger generations. The same phenomenon is going on in poorer countries, but the variance in education for current generations is still quite high. For example, in societies where a high percentage of the population works in agriculture, people often stay away from schools to help run the family farm. While this could happen in wealthier countries as well, it is more common to send the person off for an agricultural degree first. In addition, fewer people stay in the more traditional sectors where continuing the family business requires no formal education. These are plausible reasons why the effect is weaker in poorer countries and stronger in wealthier ones. The first finding presented here could have been revealed by a standard MLM analysis; the second one could have only been formally tested in an MSEM model.

Further, note that the only Level 2 covariance that is included is the one that was estimated in the random intercepts model. If theoretically relevant, one could test additional covariance components, such as the relationships between the slopes or the relationships between the intercepts and the slopes. After careful theoretical development, we could potentially hypothesize how the intercept of one of the endogenous variables (such as income) is related to the relationship between education and self-expression values when considered on the country level. While these deep theoretical considerations have been sparse even in the multilevel regression modeling literature, that did not stop the method from becoming immensely popular among social scientists. As is also the case with multilevel models, with MSEM, our empirical models allow for greater flexibility than is normally encountered in our theoretical frameworks. However, to test such a relationship, we should not forget that our sample size at the country level is still 55, and a convenience sample at that, and that we are working with variance estimates that are already unstable due to the limited sample size. These estimates are extremely underpowered and hence quite unreliable even when statistically significant results are found. Unfortunately, larger samples are inconceivable for research using cross-country surveys in the absence

Table 2.2 Model Fit Statistics

	Deviance ($-2LL$)	AIC	BIC	Parameter
Random intercepts	460,015.6	460,049.6	460,196.8	17
Random slopes	457,398.7	457,444.7	457,643.9	23

Note: Estimates from the random intercepts model are displayed in Figure 2.3, while those from the random intercepts and random slopes model are found in Figure 2.5.

of enough countries in the world but are very plausible when studying voters in precincts, students in classrooms, or patients of doctors, for example.

Comparison of Random Intercepts and Random Slopes Models

Comparing the random intercepts model with the random slopes model in Table 2.2, we get the sense that the latter represents an improvement in fit compared to the former. The deviance for the random intercepts model is 460,015.6, while for the random slopes model, it is 457,398.7. The difference between the two is 2,616.9, which is highly statistically significant when considering that the critical value for a χ^2 distribution with $23 - 17 = 6$ degrees of freedom is 12.59. This suggests that the random intercepts model fits the data significantly worse than the random slopes model. Both the AIC and the BIC reinforce this conclusion as they are considerably smaller for the latter.

Mediation and Moderation

One of the powers of MSEM over a simple multilevel regression model is our ability to test the "structure" of relationships that go beyond one dependent variable associated with multiple independent variables. Through this newly gained flexibility, we can test relationships of mediation inside our structure of associations, which opens up the potential for gauging direct and indirect effects of covariates of interest.

Mediation

A relationship of *mediation* exists in instances where an association between an exogenous covariate and an outcome can be shown to be transmitted through a third variable, the mediator. While moderation

Figure 2.6 Basic Framework for Statistical Mediation and Moderation

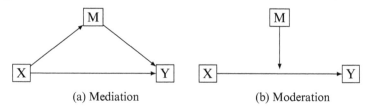

(a) Mediation (b) Moderation

explains variations in the strength of the association between a covariate and an outcome, mediation will give a precise account of how this association is transmitted, under the form of a specific *transmission mechanism* (Baron & Kenny, 1986). Figure 2.6a depicts this instance, with M acting as a mediator in the relationship between X and Y. While providing a richer description of how social phenomena unfold, mediation also relies on a set of more stringent assumptions. The pathways depicted between X, M, and Y require that the researcher defend a strict temporal ordering, with X causally prior to M and M causally prior to Y.

When discussing the simple mediation framework (Baron & Kenny, 1986) in the context of multilevel models, the wide variety of configurations that can be produced with only three variables, X, M, and Y, cause difficulties for the standard MLM setup. Even a standard configuration, such as a Level 1 mediator for the relationship between a Level 2 exogenous covariate and a Level 1 outcome, leads to problems of confounding of within-group effects (from M to Y) and between-group effects (from X to M), as pointed out by Zhang, Zyphur, and Preacher (2009). Yet possibilities abound beyond this $2 \rightarrow 1 \rightarrow 1$ situation, if we use the notation convention introduced by Krull and MacKinnon (2001), where the numbers denote the level at which a certain variable in the mediation chain is. We could have instances where a Level 1 M mediates the association between a Level 1 X and a Level 2 Y ($1 \rightarrow 1 \rightarrow 2$). These *micro-macro* effects (Snijders & Bosker, 1999), along with a variety of $2 \rightarrow 1 \rightarrow 2$, $1 \rightarrow 2 \rightarrow 2$, or even $1 \rightarrow 2 \rightarrow 1$ configurations, cannot be tested at all in a standard MLM setup, due to its inability to accommodate an outcome variable at the Level 2 (Preacher, Zyphur, & Zhang, 2010, p. 211). This is where the MSEM framework proves particularly useful.

In the example multilevel structural equation models presented, we have multiple relationships that could have both direct and indirect effects on self-expression values. One clear example is the relationship between age and self-expression values. In the multilevel regression model of Figure 2.1, we see that the impact of age on self-expression values is -0.033^{***}. The estimate based on the random intercept model in Figure 2.3 is actually identical. But Figure 2.3 highlights that the relationship between age and self-expression values can go through education, given that age and education are related. So, in addition to the direct effect of age, obtained after partialling out the influence of education and income, there also exists an indirect effect of age, transmitted through education.[2] In this specific instance, then, we are dealing with a 1 → 1 → 1 mediation setup. In a single-level SEM, we would use tracing rules to find the indirect effect of age on self-expression values through education. It would simply be the product of the coefficient from age to education and education to self-expression values (the *product-of-coefficients* method).

However, in a multilevel setting, the two random slopes coefficients are assumed to be random variables with a bivariate normal distribution, which means that applying this simple rule would lead to biased estimates (Kenny, Korchmaros, & Bolger, 2003). In practice, the expected value of an indirect effect is the multiplication of the two coefficients plus their covariance (Goodman, 1960; Kenny et al., 2003). Therefore, to find the indirect effect of age on self-expression values, we look at the product of the mean effect from age to education and education to self-expression values, summed with the covariance between the two. This rule only applies if both of the paths have a random effect.[3] The maximum likelihood estimate of this product is $\gamma_{10}^{EDU} * \gamma_{20}^{SEV} + Cov(\gamma_{10}^{EDU}, \gamma_{20}^{SEV}) = -1.427 \times (-0.204) + 0.001 = 0.293^{**}$.[4] Similarly, education has a direct effect on self-expression values ($\gamma_{20}^{SEV} = -0.204^{***}$), as well as an indirect one, through income. In this case, single-level tracing rules apply, since at least one of the paths, or in our case both education on income and income on self-expression values, is not allowed to vary across countries. Therefore, $\gamma_{10}^{INC} \gamma_{10}^{SEV} = 0.015^{***}$.

---

[2] We also theorized an indirect effect through income, but the results revealed that age does not explain income, at least in our model specification.
[3] The implementation of this formula in various software packages is given by Bauer, Preacher, and Gil (2006).
[4] The standard error–based confidence intervals for this product can be obtained directly through maximum likelihood estimation using Sobel's (1982, 1986) delta method (Bollen, 1987)—not to be confused with the delta parameterization available in *Mplus* for categorical outcomes. In small samples, this estimate may be biased, in which case a bootstrap is preferable (Preacher et al., 2010).

The model we test, though, presents an additional mediation: *GDP*, a Level 2 exogenous covariate, exerts its effect on lower-level variables (like education and self-expression values) both directly and indirectly through its effect on a Level 1 covariate of self-expression values: education. This is a 2 → 1 → 1 configuration. In the MLM setting, the direct effect of *GDP* on self-expression values is transmitted through the intercept of *SEV*, which is allowed to vary across the countries.

Figure 2.7 extracts just the parts of the model that are of interest when assessing the direct and indirect effect of *GDP* on self-expression values: the slope of *EDU* on *SEV* and how this slope, as well as the intercepts of *EDU* and *SEV*, are affected by *GDP*. Here, the line separating the levels of analysis is now vertical and the mediation is presented in the usual triangular form. At the individual level, the effect of GDP per capita on self-expression values is clearly moderated by educational achievement ($\gamma_{21}^{SEV} = 0.027^{***}$). At low levels of GDP per capita, the gap between those with high and low education is *larger* than the corresponding gap in countries with high levels of GDP per capita. At the same time, the impact of *GDP* on self-expression values confirms what we discovered in our random intercepts specification: Wealthier countries exhibit higher levels of self-expression values in the citizenry.

Note how, as opposed to the model in Figure 2.3, in Figures 2.5 and 2.7, *GDP* has a significant impact on education ($\gamma_{01}^{EDU} = -0.289^{*}$). This significant path is what makes the assessment of a potential indirect effect of GDP per capita on self-expression values worthwhile. Once again, just like in single-level SEM, the maximum likelihood estimate of the indirect effect of *GDP* on *SEV* is the product of the two direct paths: from *GDP* to *EDU* and from *EDU* to *SEV*. In this case, this yields a significant positive effect: $\gamma_{01}^{EDU} \times \gamma_{20}^{SEV} = -0.289 \times -0.203 = 0.059^{**}$, which is depicted in Figure 2.7 using a dotted arrow.

Moderation

Unlike mediation, *moderation* refers to an instance where a variable alters the strength or direction of the relationship between an exogenous and an endogenous variable (Baron & Kenny, 1986, p. 1174), as is depicted in Figure 2.6b. In this framework, there is a debate on the causal ordering of X, M, and Y. Kraemer et al. (2008; 2002) argue that M must be prior to and uncorrelated with X. For this reason, for example, the same variable could not be a mediator and moderator at the same time, since a mediator M is necessarily directly associated with

Figure 2.7 Moderated Mediation Effect in the Slopes Model

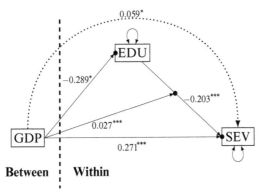

Note: Figure includes only the variables involved in the effect of *GDP* on self-expression values, both directly and through education. The line separating the levels of analysis is now vertical and the mediation is presented in the usual triangular form.

X by means of a causal pathway. Hayes (2013, pp. 399–402), however, demonstrates mathematically how this independence between M and X is not a requirement, and therefore, the causal link between X or Y and M could be modeled if it makes theoretical and substantive sense. Hayes (2013, pp. 209–210) lists a large number of theories in social sciences that rely on evidence provided by moderation. Petty and Cacioppo's (1986) *elaboration likelihood model*, for example, relies extensively on arguments that imply moderation. Persuasion effects will be stronger depending not only on characteristics of the message (e.g., if it originates with an expert) but also on an individual's motivation and ability to think about the specific topic that the message addresses. In their account, individual motivation moderates the association between message characteristics and persuasion. Solt's (2008) updated version of Goodin and Dryzek's (1980) *relative power theory* sees income inequality as moderating the relationship between income and political participation: In countries with higher inequality, the difference in terms of participation between rich and poor voters is higher than in countries with lower-income inequality.

In the model we test here and which we present in a truncated form in Figure 2.7, GDP per capita has a direct effect on education and self-expression values but also moderates the relationship between education and self-expression values. In countries with high GDP (or, in this case, the natural logarithm of GDP), the negative relationship between education and self-expression values becomes weaker. This represents precisely the type of moderated relationship we have discussed above, which takes place between a Level 2 moderator and a Level 1 relationship. A second mediation relationship is present in Figure 2.5, although not also depicted in Figure 2.7: The impact of age, mediated by education, is now also moderated by GDP.[5]

In essence, the random slopes model gives us the ability to test both direct and indirect relationships moderated by higher-level characteristics. In addition to this between-level moderation of within-level relationships, it is possible to assess direct and indirect effects of a higher-level (between-level) phenomenon on a lower-level (within-level) outcome. We have seen from Figure 2.5 how, in the presence of random slopes, it is somewhat difficult to read within-level direct and indirect effects. One has to consult the structure in both the within and the between level. When aiming to assess direct and indirect effects across levels, the situation is quite similar.

In the random slopes model, these sets of relationships become more complex. The first issue that is apparent is that between Figures 2.3 and 2.5, the sign of the direct effect of age on self-expression values has changed. This may not be apparent at first glance in Figure 2.5 as this relationship is now allowed to vary across the countries, making it a latent variable in the between level. Still, this latent's mean expressed on the path going from the triangle to β_{3j}^{SEV} highlights how this relationship is now positive and significant and not negative and significant. The reasons for this change are manifold. First, this model is substantially different and, based on the comparison of fit, better describes the data. Some of the indirect effects are now modeled differently. More important, though, in the second model, an interaction is modeled explicitly. The impact of age on self-expression values, now moderated by GDP,

---

[5] In the interest of clarity, as well as to assist the reader in absorbing these concepts, we have chosen to present these two concepts as separate. In our model, though, the connections between GDP, education, and self-expression values represent a fairly typical case of moderated mediation across two levels of analysis: country and individual. In SEM, moderated mediation is typically modeled on a single level of analysis through the inclusion of the product of the moderator and the explanatory factor of interest, which is included as a new variable in the model. This is not the case here. The product variable is not explicitly included as an additional variable; rather, the moderation is modeled as the impact of a between variable on the random effect of a path.

decreases with every doubling of GDP. Considering the range of GDP's logarithm (between 6.99 and 11.23 in this data set), this interaction does not fully account for the change here, but it is true that for the United States, this direct impact (which, we should not forget, is also controlling for indirect impacts) is estimated to be negative.

Centering in MSEM

The reader might be puzzled as to why we did not use a common technique in the multilevel modeling world for easing the interpretation of this interaction: centering. Widely discussed in the multilevel modeling literature (Enders & Tofighi, 2007; Kreft, de Leeuw, & Aiken, 1995; Paccagnella, 2006), centering is also applicable in the case of multilevel structural equation models. To produce a more meaningful interpretation of coefficients, centering can take the form of either group mean or grand mean centering (see Enders & Tofighi, 2007). If *GDP* were grand mean centered, the direct effect of age on self-expression values would be estimated for the average GDP level (as *GDP* would be 0 for the country with the average GDP). Group mean centering of Level 1 variables, on the other hand, changes their interpretation as well as their estimated effect on a Level 1 outcome. This occurs because group mean centering (subtracting the average for each country out of the measured values on an indicator such as age or income) erases all between-country variation in a variable. In the aftermath of this procedure, all that is preserved is relative positions of Level 1 units inside a group. Although the average age in Japan is much higher than the average age in Egypt, after centering, only relative differences between individuals inside a country would be preserved. This produces a clear estimate of a Level 1 variable's effect on a Level 1 outcome, disregarding all dynamics between countries.

In MSEM, centering is complicated by the possibility of statistical associations between indicators measured at various levels of the data hierarchy. MSEM can not only accommodate the standard path modeling situation (1 → 1 → 1), but it can also estimate more intricate pathways of association: 2 → 1 → 1, 1 → 2 → 2, or even 1 → 2 → 1. This added flexibility comes at the cost of greater care when deciding whether a relationship in the pathway should be estimated based on within or between variation.

We begin our discussion by relying on a framework proposed by B. O. Muthén and Asparouhov (2008). Multilevel mediation can be estimated by partitioning the multilevel structural equation model into (1) a measurement model, (2) a "within" structural model, and (3) a "between" structural model.[6] To achieve this separation of the structural model, each Level 1 observed indicator is partitioned into a "within" and a "between" latent component. These are then used to specify the (2) and (3) models: A 2 → 1 component would be estimated using only the "between" latent of the Level 1 indicator, while the continuing 1 → 1 pathway would be estimated using only the "within" latent of the same indicator.

In *Mplus*, these steps are performed automatically as part of the estimation procedure, even if the data contain raw (before centering) versions of the indicators (Preacher et al., 2010, p. 215).[7] The reader should therefore be confident that centering was performed in the analyses we report so far, even though we have not discussed centering until now. Taking the example of the model presented in Figure 2.5, the "within" effect of education on self-expression values is estimated using only "within" variation in education, as *Mplus* has already performed the decomposition into two latents we discuss above. On the other hand, the "between" effect of GDP per capita on the intercepts for education is estimated using only "between" variation in education.

Other software does not automate this partitioning process, which is why the researcher is entrusted to perform it. In such instances, a researcher would construct the group-level latent *configurational* construct (using the terminology introduced by Kozlowski & Klein, 2000) out of the individual-level observed indicators. Following this, the researcher would proceed as in standard MLM by making sure that any 1 → 1 pathway is estimated using the "within" latent and that a 2 → 1 or 1 → 2 pathway, or even the more complicated patterns found in three-level MSEM, is estimated using the "between" (*configurational*) latent.

Summary

This chapter highlighted examples that are possible when path analyses are extended into the multilevel framework. The kind of modeling flexibility demonstrated here is simply not possible using a basic multilevel

---

[6] We encourage more advanced readers to consult B. O. Muthén and Asparouhov's original text or, at a minimum, the brief exposition made by Preacher et al. (2010).

[7] See also the original discussion in L. K. Muthén and Muthén (1998–2017, p. 261) related to group mean centering when disentangling "within" and "between" variance components for latent covariate decomposition.

model and, in fact, goes beyond what we have shown so far. For example, it is entirely possible to model full structures of relationships on the between level. The models presented here used a single Level 2 variable, *GDP*, in both the context of moderating Level 1 relationships and as a crucial covariate of a Level 1 phenomenon. But it is possible to include multiple Level 2 variables in the model and not just as exogenous variables. It would be possible to include other country-level phenomena not just as direct predictors of Level 1 outcomes and moderators of Level 1 relationships but also as predictors of Level 2 phenomena, extending the possibility of cross-level mediation structures. Say, for example, *GDP* was the mediator of a Level 2 explanatory factor that had both direct and indirect effect on self-expression values. The number of modeling possibilities seem to be limited only by researchers' theories and available data.

While latent variables have already emerged on Level 2, as varying intercepts and slopes, the real flexibility of SEM, in its single-level form, comes from its ability to include multiple-indicator latent variables. In the following chapter, we present the most basic latent variable SEM, a confirmatory factor model's extension into multilevel modeling, as a stepping-stone to full structural models in the multilevel framework.

CHAPTER 3. MULTILEVEL FACTOR MODELS

The previous chapter highlighted how multilevel and path models can be combined to strengthen each other's typical features by allowing the modeling of complex relationships on multiple levels of analysis. A second advantage of the SEM framework is its ability to incorporate better developed measurement models into path analysis. Techniques such as Seemingly Unrelated Regression can already handle multiple observed outcomes but have a harder time dealing with clustered data. When confronted with a battery of items tapping into the same phenomenon, applied researchers most often resort to constructing an average of the battery in an ad hoc way. In the SEM framework, the same process can be pursued more rigorously by means of confirmatory factor analysis (CFA), in which the researcher imposes a measurement structure to the data.[1] It involves considering a number of observed variables as manifestations of one or more latent, unobserved variables. This chapter is dedicated to multilevel measurement models, focusing on two-level CFA.

For the examples in this chapter, we use data from the 2015 wave of the Programme for International Student Assessment (PISA) for the Dominican Republic. The PISA is an Organisation for Economic Co-operation and Development (OECD)–sponsored cross-national survey of 15-year-old students' performance in science, reading, and mathematics (in the most recent wave, collaborative problem solving is also assessed). Besides educational performance test scores, it also includes surveys with students, parents, teachers, and school principals. PISA offers a classic example of multilevel modeling in education studies, where students can be clustered in schools and schools in countries. In PISA, schools are sampled proportional to size (PPS) in each country, and students are randomly sampled within each school. We use data only from the Dominican Republic to make examples easier to follow—the hierarchical structure after listwise deletion involves 3,203 students clustered in 186 schools. Due to the students coming from different schools, this is a two-level data structure with a relatively large Level 2 sample and an average cluster size of around 17, making it appropriate for multilevel analysis.

---

[1] Multilevel exploratory factor analysis (EFA) is also possible. However, due to space limitations, we constrain our discussion to confirmatory factor analysis, which is later integrated into path models. Interested readers on multilevel EFA can consult van de Vijver and Poortinga (2002).

Let's take, for example, the issue of students' Internet use. In our case, we are particularly interested in various educational and recreational uses of the World Wide Web. Normally, we would tap such a construct with multiple survey questions and then test its dimensionality and the items' effectiveness through factor-analytic techniques. However, just like with regression and path models in the previous chapter, these factor-analytic approaches assume independent observations, meaning that the clustering across schools could hinder our ability to obtain unbiased results and appropriate model fit. Furthermore, we could be interested in how much difference there is in Internet use across students from different schools and on whether the kind and amount of use are different across schools. To address these and other potential questions, we proceed by generalizing multilevel models to confirmatory factor analysis.

We use six indicators from the student survey section of PISA in the examples. All indicators deal with the use frequency of electronic devices and the Internet outside of school time. A full list is in Table 3.1. They are all measured on a 1-to-5 scale, from *never or hardly ever* used to used

Table 3.1 List of Questions and Variables Acronyms

Code	Question
Use of digital devices for leisure (FUN)	
VID	Use digital devices outside school for browsing the Internet for fun videos (e.g., YouTube)
DL1	Use digital devices outside school for downloading music, films, games, or software from the Internet
DL2	Use digital devices outside school for downloading new apps on a mobile device
Use of digital devices for school work (SCH)	
ISC	Frequency of use outside of school: browsing the Internet for schoolwork (e.g., for preparing an essay or presentation)
ILS	Frequency of use outside of school: browsing the Internet to follow up lessons (e.g., for finding explanations)
HWC	Frequency of use outside of school: doing homework on a computer

Note: All questions asked with the following 1 to 5 response scale: 1 = *never or hardly ever*, 2 = *once or twice a month*, 3 = *once or twice a week*, 4 = *almost every day*, and 5 = *every day*.

Figure 3.1 Basic CFA Model of Digital Devices Use

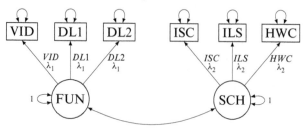

every day, and are treated here as continuous for the sake of simplicity.[2] We hypothesize that these indicators measure two kinds of Internet use: The first three are connected to uses for leisure, while the latter three are uses for schoolwork. Therefore, the initial CFA measurement model is described in Figure 3.1.

The model in Figure 3.1, even if estimated without taking the clustering of observations into account, fits well to the data. RMSEA is a bit high, at .068, but other fit indices suggest acceptable fit: CFI is .987, TLI is .976, and SRMR is .029. All standardized factor loadings are above .65. The correlation between the latent variables is moderate, at .645—not high enough to suggest that in fact we have only one underlying dimension.

This example shows that the move from single-level to multilevel CFA has to be justified substantively, given the hierarchical nature of a data set, and not based on model fit. Ignoring the clustering of students into schools violates the assumption of independent observations, but this does not prevent the model from showing good fit. Besides taking care of violated assumptions, we might have theoretical considerations in mind about different relationships between the measurement variables across schools that justify going for a multilevel model. For example, while it may be meaningful to talk about uses of digital devices for different purposes among students, it is possible that across schools, the

---

[2] The aim of this example is to demonstrate the method and not to test substantive theories. Weights, complex missing data treatments, and ordinal variables are not used in any of the examples. Originally, this data set contained 3,857 students from 192 schools. PISA suggests weights to be included in the analysis and that public and private schools be analyzed separately. The issue of weighting, limited endogenous variables, and more sophisticated missing data treatments are discussed in the final chapter. Also, researchers who wish to account for the difference between private and public schools can use multigroup multilevel CFA. All of these issues are described in more detail in the final chapter of this book.

difference disappears: In some schools, students use more Internet on average for all purposes, and in others, they use less. That kind of hypothesis can be tested with multilevel factor models, as we demonstrate later in the chapter.

Confirmatory Factor Analysis in Multiple Groups

The first way to deal with clustering in CFA, and SEM more generally, is through multiple group analysis. For example, if a researcher suspects that a few model parameters might behave differently between male and female participants, she can use a multiple group model and allow those parameters to vary between groups.[3] Similarly, if one has survey respondents from different countries, it is possible that indicators, as well as their relationships to one another, vary from country to country.

In the case of measurement models that have multiple-indicator latent variables, the property of indicators measuring the same construct in the same way across different groups is called *measurement invariance*. An instrument is invariant when two individuals who have the same level of the latent construct, from two different populations, would have the same score on the observed indicator (say, give the same answer on a survey item). It is *noninvariant* if those two individuals would have different scores on the observed indicator because of their different group membership.

The literature identifies four levels of (non)invariant CFA models (Meredith, 1993): *Configural* is the least restrictive, in which all parameter estimates are allowed to vary across groups. *Metric* invariance models constrain factor loadings to be equal across groups. *Scalar* invariance requires that both factor loadings and indicators' intercepts are the same across groups, and *strict* invariance imposes equality constraints across groups upon factor loadings, indicators' intercepts, and the indicators' error terms. If the measurement instrument fails a scalar invariance test, direct comparison of group factor means is not possible, since we cannot rule out it being caused by differential functioning of a given item across groups. However, for regression purposes, it is usually enough to achieve *metric* invariance across groups.

Multiple-group CFA (MGCFA) can give answers on measurement invariance and also hypothesis tests about measurement models. For

---

[3] Not only on the measurement part. Multigroup SEM also tests differences in structural components of the models across groups, such as regression coefficients.

example, suppose there are some schools in areas where the most digital devices available are smartphones. Students interested in class content might still use smartphones for preparing for classes or following up on lessons, but they are less likely to use them for doing homework. Therefore, in those groups, HWC would likely be a bad indicator of usage of digital devices for schoolwork. MGCFA would be one way of testing this hypothesis.

While MGCFA is a good point to start thinking of CFA with hierarchical data, multilevel CFA extends the possibilities of modeling and analysis in a few ways. First, conceptually, it not only incorporates variations in measurement across groups but also allows the testing of different measurement structures across the two (or more) levels—for instance, to test whether indeed the same structure of use for fun versus use for schoolwork is present across schools as it is across students. Second, one can add Level 2 covariates to explain variance at Level 2 constructs and expand the model into a full structural regression (as seen in Chapter 4). And third, it is possible to perform measurement invariance tests across many groups by allowing factor loadings to vary across clusters, which is more practical with a large number of clusters than the traditional multiple-group CFA invariance test. This is discussed in the last section of this chapter.

Two-Level CFA

Multilevel CFA follows the estimation of multilevel path models in splitting the variance of each variable into within- and between-group variance components, which are additive and uncorrelated (B. O. Muthén, 1989, 1994). This means that the total variance of an observed variable Y is the sum of its within and between variances, or $Var(Y) = Var(Y)_B + Var(Y)_W$. This is the same type of variance partitioning encountered in random intercept baseline models in multilevel modeling. To define a single variance component for the within part, it is necessary to assume that the relationships between variables are the same within each group. In other words, the covariance matrix of group g_1 is the same as that of groups $g_2, g_3, \ldots g_j$ (Goldstein & McDonald, 1988; McDonald & Goldstein, 1989). This marks the first difference when comparing to multiple-group CFA; the latter focuses on testing whether the relationships between variables change from one or a few specific groups to the others. Put another way, two-level CFA without random loadings assumes measurement invariance.

This division is made clear by considering the set of equations that defines a two-level CFA. Here we expand the example from Section 3.1, in which the Level 1 (students) factor structure is reproduced at the Level 2 (schools). We start with the within part of the model in Equation 3.1:

$$\begin{cases} VID_{ij} = \lambda_{0j}^{VID} + \lambda_1^{VID} FUNW_{ij} + \varepsilon_{ij}^{VID} \\ \\ DL1_{ij} = \lambda_{0j}^{DL1} + \lambda_1^{DL1} FUNW_{ij} + \varepsilon_{ij}^{DL1} \\ \\ DL2_{ij} = \lambda_{0j}^{DL2} + \lambda_1^{DL2} FUNW_{ij} + \varepsilon_{ij}^{DL2} \\ \\ ISC_{ij} = \lambda_{0j}^{ISC} + \lambda_2^{ISC} SCHW_{ij} + \varepsilon_{ij}^{ISC} \\ \\ ILS_{ij} = \lambda_{0j}^{ILS} + \lambda_2^{ILS} SCHW_{ij} + \varepsilon_{ij}^{ILS} \\ \\ HWC_{ij} = \lambda_{0j}^{HWC} + \lambda_2^{HWC} SCHW_{ij} + \varepsilon_{ij}^{HWC} \end{cases} \quad (3.1)$$

The between-level part, taking the observed variables' group intercepts (λ_{0j}) as indicators, is described in Equation 3.2. And both can be merged into a single set, as in Equation 3.3. The variance of each observed indicator is decomposed into five elements. First is an overall intercept across groups (μ_{00}), or the *grand mean*. Next is the contribution for the within-level variance, divided into what is accounted for by the latent variable ($\lambda_1 FUNW_{ij}$ or $\lambda_2 SCHW_{ij}$) and the residual, unexplained within-level variance (ε_{ij}). The between-variance is denoted by the part explained by the between-level factor ($\mu_{01} FUNB_j$ or $\mu_{02} SCHB_j$) and the residual, unexplained between-level variance (υ_{0j}).

$$\begin{cases} \lambda_{0j}^{VID} = \mu_{00}^{VID} + \mu_{01}^{VID} FUNB_j + \upsilon_{0j}^{VID} \\ \\ \lambda_{0j}^{DL1} = \mu_{00}^{DL1} + \mu_{01}^{DL1} FUNB_j + \upsilon_{0j}^{DL1} \\ \\ \lambda_{0j}^{DL2} = \mu_{00}^{DL2} + \mu_{01}^{DL2} FUNB_j + \upsilon_{0j}^{DL2} \\ \\ \lambda_{0j}^{ISC} = \mu_{00}^{ISC} + \mu_{02}^{ISC} SCHB_j + \upsilon_{0j}^{ISC} \\ \\ \lambda_{0j}^{ILS} = \mu_{00}^{ILS} + \mu_{02}^{ILS} SCHB_j + \upsilon_{0j}^{ILS} \\ \\ \lambda_{0j}^{HWC} = \mu_{00}^{HWC} + \mu_{02}^{HWC} SCHB_j + \upsilon_{0j}^{HWC} \end{cases} \quad (3.2)$$

$$\begin{cases} VID_{ij} = \mu_{00}^{VID} + \lambda_1^{VID} FUNW_{ij} + \mu_{01}^{VID} FUNB_j + v_{0j}^{VID} + \varepsilon_{ij}^{VID} \\ DL1_{ij} = \mu_{00}^{DL1} + \lambda_1^{DL1} FUNW_{ij} + \mu_{01}^{DL1} FUNB_j + v_{0j}^{DL1} + \varepsilon_{ij}^{DL1} \\ DL2_{ij} = \mu_{00}^{DL2} + \lambda_1^{DL2} FUNW_{ij} + \mu_{01}^{DL2} FUNB_j + v_{0j}^{DL2} + \varepsilon_{ij}^{DL2} \\ ISC_{ij} = \mu_{00}^{ISC} + \lambda_2^{ISC} SCHW_{ij} + \mu_{02}^{ISC} SCHB_j + v_{0j}^{ISC} + \varepsilon_{ij}^{ISC} \\ ILS_{ij} = \mu_{00}^{ILS} + \lambda_2^{ILS} SCHW_{ij} + \mu_{02}^{ILS} SCHB_j + v_{0j}^{ILS} + \varepsilon_{ij}^{ILS} \\ HWC_{ij} = \mu_{00}^{HWC} + \lambda_2^{HWC} SCHW_{ij} + \mu_{02}^{HWC} SCHB_j + v_{0j}^{HWC} + \varepsilon_{ij}^{HWC} \end{cases}$$

(3.3)

Estimation

Following a common practice of multilevel modeling, the first step in an analysis that involves hierarchical data is inspecting the intraclass correlation coefficients (ICCs) for each of the indicators: *VID*, *DL1*, *DL2*, *ISC*, *ILS*, and *HWC*.[4] In the PISA data used for this chapter, the estimated ICCs for indicators range from .046 (*DL2*) to .149 (*VID*), with the others around .10.

From this point onward, two approaches may be followed. Hox (2010, pp. 300–301) suggests following the standard practice in multilevel modeling: Start with a null model (with no substantive covariates of the outcome at either level of the model). Second, fit an *independence model* that estimates Level 2 variances for each indicator but no Level 2 covariances. And third, add a covariance component at the between level by placing no restriction on the estimated Level 2 variances or covariances. This stepwise approach gives a further test as to whether there is between-group variance to be explained (beyond the ICC) and if this variance is structural and not only the result of random sampling variation.

We suggest a more theory-driven approach, which follows more closely the SEM tradition: Once the existence of between-level variance

---

[4] The ICC is expressed as the share of between-group variance out of total variance in the outcome (Snijders & Bosker, 1999, pp. 16–22). As for any correlation, it ranges from 0 to 1, with higher values denoting that groups are increasingly different from each other and more homogeneous internally. In a multilevel setting, we ideally want higher ICC values, as these denote there is substantial between-group variance in the outcome, to be explained with group-level covariates.

is identified, through the ICC, a researcher fits the model(s) reproducing her theoretical expectations about the data. Therefore, in the first two-level CFA, we test the same factor structure on both levels, following Equation 3.3: The latent construct "use of digital devices for leisure" is measured by three indicators, while "use for schoolwork" is measured by another three. This structure is the same at both the within level (students) and the between level (schools). The model is depicted in graphical format in Figure 3.2. The Level 2 part of the model takes the estimated intercepts of the items for each school as indicators.[5] This is denoted by the solid dark dots at the end of arrows in the within part and by the circles, latents, in Figure 3.2. The Level 2 indicators are parameters estimated at the Level 1. In this depiction, we allow all factor loadings to be estimated and fix factor variances to 1. It is possible to use the alternative identification approach: Fix one of the loadings for a latent variable to 1, and allow the factor variance to be estimated.

Identification

To calculate the number of free parameters that one can include in a multilevel SEM, we slightly modify the formula $p(p + 1)/2$, described in Chapter 1, in which p is the number of observed variables. Since now we have two covariance matrices (within and between), we have double the number of entries, and therefore the formula is multiplied by 2 (or, rather, not divided by 2). Furthermore, we add the number of observed variable means, that is, indicators' intercepts. In short, the maximum number of free parameters for the global model is given by $p(p + 1) + k$, where k is the number of indicator intercepts (Heck & Thomas, 2015, p. 165).

For the model in Figure 3.2, there are six observed variables (p), each one with an estimated indicator (k). Replacing that into the identification formula, we find that $6(6 + 1) + 6 = 48$ free parameters can be estimated in the global model. The number of parameters estimated at the within level is six residuals, six factor loadings, and one factor covariance. The factor variances are fixed to one in order to set the latent variables' metric. In total, there are 19 estimates. For the between level, we have the six intercepts,[6] six residuals, six factor loadings, and one factor covariance, summing up to 13. Hence, the global model has a total of 32 parameters estimated. With a maximum allowed of 48, the model we described earlier is overidentified and can be estimated.

---

[5] This is why we should never group mean center indicators in multilevel CFA. Such a move would fix the between-group variance of intercepts at zero.

[6] Intercepts are estimated only at one level.

Figure 3.2 Two-Level CFA Model of Digital Devices Usage

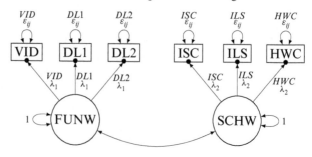

Results

Results for the model from Figure 3.2 are displayed in Figure 3.3. We find a model with better fit than the one-level CFA presented earlier: While the χ^2 test is still significant at $p < .001$, RMSEA is now below .05, at .046. CFI and TLI remain the same, .985 and .973, both of which still indicate good fit. SRMR indicates good fit on both levels: It is .03 for within and .028 for between. Factor loadings are high on both levels, indicating a reliable measurement instrument.[7] However, the correlation between the two school-level factors is .939 (the .269 in the figure is their

---

[7] To maintain consistency with all models in the book, we present unstandardized loadings.

covariance). Such a high correlation suggests that, at the school level, all six indicators might be measuring a single construct.

Partial Saturation Fit Test

Global fit indices suggest a good fit for the model in Figure 3.3. However, by assessing the fit of the entire model, these can hide lack of fit particular to one of the levels. Ryu and West (2009) suggest a way of dealing with this problem: the partially saturated model fit evaluation, hinted at in Chapter 1. It works by refitting the model in two ways: first with the hypothesized Level 1 model and a saturated Level 2 model (PS_W)—this means that at Level 2, all variables are left free to correlate with one another and no model is imposed. Second, fitting the hypothesized Level 2 model, leaving all variables at the Level 1 free to correlate with

Figure 3.3 Two-Level CFA Model of Digital Devices Use

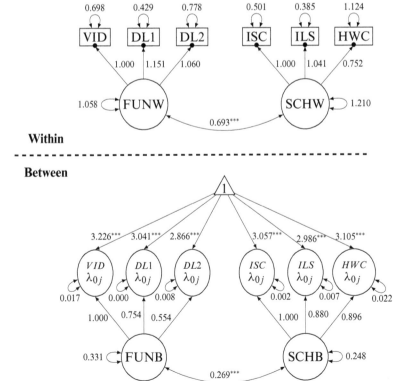

one another without any model imposed on them (PS_B). Saturated models have perfect fit to the data. Therefore, for the PS_W model, in which the Level 2 is saturated, any lack of fit obtained by the χ^2 test (and its derivatives, such as CFI and RMSEA) is attributable to the within level. Conversely, in the PS_B model, where the Level 1 is saturated, any lack of fit can be attributed to the between level.

When running the PS_W model for Figure 3.3, we find a $\chi^2 = 109.589, df = 8, p < .001$, RMSEA of .063, CFI of .986, TLI at .949, and SRMR of .030. These numbers are close to and sometimes worse than the global fit obtained for Figure 3.3, which suggests that most of the lack of fit in that model comes from the within part. Running the partially saturated model to test the between-level fit, however, gives an interesting result. While model fit is good, we run into an estimation problem due to multicollinearity, and one estimated residual variance is negative ($DL1$). This issue, along with the .939 correlation between the two Level 2 factors, further indicates that the factor structure at the between level might be misspecified. Judging by the factor correlation, it is probably unidimensional, formed by all indicators.

Unidimensional Level 2 Factor Structure

Results of the model with a unidimensional Level 2 factor structure are in Figure 3.4. They show virtually the same global fit as that of Figure 3.3: a significant χ^2 test, RMSEA at .047, and almost the same CFI (.984) and TLI (.971 against .973 in the previous). The main difference is that SRMR indicates a slightly worse fit of the between part, at .055 (against .028 in the previous), while the SRMR within remains close, at .032. All factor loadings at the school level are high, indicating they do covary on a single dimension. Using the partial saturation approach, however, reveals that the multicollinearity issue at the Level 2 disappears: The estimation runs without a problem, and the Level 2 specific fit indices are $\chi^2 = 42.209, df = 9, p < .001$, RMSEA = .034, CFI = .996, TLI = .985, and SRMR between = .055.

The stepwise approach taken here illustrates, first, that the decision to run a two-level CFA instead of a one-level CFA has to be grounded on theory and on the data structure, not purely on model fit. The one-level CFA model had good fit to the data. Second, while a two-level CFA may show good global fit, it is important to test the fit of each level separately to check if the global test statistics are not hiding level-specific lack of fit. In this case, the test proposed by Ryu and West (2009) showed that the Level 2 is better specified as a single-factor model rather than as a repetition of the two-factor model from Level 1.

Figure 3.4 Two-Level CFA Model of Digital Devices Use With a Unidimensional Level 2 Model

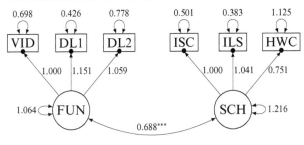

Conceptually, we can interpret these findings in the following way: While for students, it makes sense to differentiate between usage of devices for fun or schoolwork, across schools, the average frequency of use for any purposes might be more an indicator of something undifferentiated by the purpose of the device (most likely wealth). Schools with students from wealthier families, who have more access to digital devices, have a higher average frequency of use for all purposes in comparison to schools with pupils from poorer families or perhaps more isolated rural areas. Digital usage, at the school level, is therefore unidimensional, even if the factor structure at the level of the person suggests two separate constructs. To test this hypothesis, a next step would be to add measures of household income and check if it helps explain the variation of devices usage across schools. This would lead us closer to a full multilevel structural model, demonstrated in Chapter 4 with another data set.

This example highlights that the factor structure at the between level does not have to be the same as that of the within level. They explain two separate variance components, and models that make conceptual and theoretical sense at Level 1 may not do so at Level 2. Although, in this case, we arrived at the more parsimonious model by looking at the between-level factor correlation, this should not stop anyone from thinking theoretically and developing models with different factor structures on the two levels of analyses.

Random Latent Variable Intercepts

As mentioned earlier, the most common topic in measurement invariance is whether a set of measurements from a battery of items is tapping into the same latent construct across groups. However, in a multilevel CFA, another kind of invariance comes into play: *cross-level invariance*. It refers to a measurement instrument that may not work the same way across the various groups but, indeed, functions the same across the two (or more) levels of analysis. In other words, the factor loading of a given indicator is the same at both the within and the between levels of the model. Much like invariance across groups is required to compare means and regression coefficients, cross-level invariance is required if we want to compare regression coefficients across the levels of analysis (Marsh et al., 2009). For example, if we want to see if the correlation between use of digital devices for fun and for schoolwork is stronger at the individual level than at the school level, we would need to have a model with cross-level invariance. Only then do the latents have the same metric on both levels of analysis and hence would be directly comparable.

By imposing an invariant measurement model, we also get an indication of whether the Level 1 latents show variation across groups. For instance, in the previous section, we find two meaningful constructs on the use of phones and computers at the student level but only one construct at the school level. However, we have not tested if there is meaningful variation across schools for students' usage of devices for fun. Say, for example, that we would simply like to test if students in some schools use devices for fun much more than students in other schools. That can only be done by estimating the variance on the average use for fun (i.e., the factor mean) across schools—which is, in effect, a random intercept model in which the intercept (mean) of the latent variable has between-group variance.

In the models presented until now, the within-level factors have no cross-cluster variance components. This is because the variation cumulates at the level of the item parameters as we allow the factors on the two levels to have different scales. The variance we observe at the school level is for the factor estimated at that level with school-level indicators (the cluster means of each observed indicator). They are a school characteristic and not across-group variation of a student characteristic.

To split the variance of within-level factors into within and between components, we need to ensure that item-level differences across the levels do not explain away this variation before we would allow it to emerge. The most obvious way to do this is to ensure that the latent variables across the two levels have a common scale by imposing equality constraints on the estimated factor loadings at both levels. More specifically, we need to build a model where the factor structure is the same across the two levels of analysis and where factor loadings for each indicator on its respective latent variable are also equal across the two levels of analyses. The equality constraint will identify this model by allowing us to disentangle across-group variances on the item level from the across-group variances on the latent level. It is important to notice that we still estimate different variances for each within- and between-level latent variable and that the covariance between the two factors does not need to be the same for Level 1 and Level 2.

For example, departing from Figures 3.2 and 3.3, the loading of $DL1$ on the latent variable $FUNW$ is fixed to be the same as the loading of λ_{0j}^{DL1} on the latent variable $FUNB$. Or more formally, the following equality constraints are imposed:

$$\begin{cases} \lambda_1^{VID} = \mu_{01}^{VID} \\ \lambda_1^{DL1} = \mu_{01}^{DL1} \\ \lambda_1^{DL2} = \mu_{01}^{DL2} \\ \lambda_2^{ISC} = \mu_{02}^{ISC} \\ \lambda_2^{ILS} = \mu_{02}^{ILS} \\ \lambda_2^{HWC} = \mu_{02}^{HWC} \end{cases} \quad (3.4)$$

By doing so, the scale of latent variables in both levels becomes the same, and their variances can be directly compared and added

(Mehta & Neale, 2005). The sum produces an individual-level latent variable whose variance can be decomposed into within- and between-cluster variances: $Var(FUN)_{ij} = Var(FUNW)_{ij} + Var(FUNB)_j$. Since $\lambda_1 = \mu_{01}$, we follow Mehta and Neale (2005, p. 273) in simplifying Equation 3.3 to produce the specification seen in Equation 3.5.

$$\begin{cases} VID_{ij} = \mu_{00}^{VID} + \mu_{01}^{VID}(FUNW_{ij} + FUNB_j) + \upsilon_{0j}^{VID} + \varepsilon_{ij}^{DL1} \\ DL1_{ij} = \mu_{00}^{DL1} + \mu_{01}^{DL1}(FUNW_{ij} + FUNB_j) + \upsilon_{0j}^{DL1} + \varepsilon_{ij}^{DL1} \\ DL2_{ij} = \mu_{00}^{DL2} + \mu_{01}^{DL2}(FUNW_{ij} + FUNB_j) + \upsilon_{0j}^{DL2} + \varepsilon_{ij}^{DL2} \\ ISC_{ij} = \mu_{00}^{ISC} + \mu_{02}^{ISC}(SCHW_{ij} + SCHB_j) + \upsilon_{0j}^{ISC} + \varepsilon_{ij}^{ISC} \\ ILS_{ij} = \mu_{00}^{ILS} + \mu_{02}^{ILS}(SCHW_{ij} + SCHB_j) + \upsilon_{0j}^{ILS} + \varepsilon_{ij}^{ILS} \\ HWC_{ij} = \mu_{00}^{HWC} + \mu_{02}^{HWC}(SCHW_{ij} + SCHB_j) + \upsilon_{0j}^{HWC} + \varepsilon_{ij}^{HWC} \end{cases} \quad (3.5)$$

In turn, this can be further simplified to produce the specification presented in Equation 3.6.

$$\begin{cases} VID_{ij} = \mu_{00}^{VID} + \mu_{01}^{VID}FUN_{ij} + \upsilon_{0j}^{VID} + \varepsilon_{ij}^{DL1} \\ DL1_{ij} = \mu_{00}^{DL1} + \mu_{01}^{DL1}FUN_{ij} + \upsilon_{0j}^{DL1} + \varepsilon_{ij}^{DL1} \\ DL2_{ij} = \mu_{00}^{DL2} + \mu_{01}^{DL2}FUN_{ij} + \upsilon_{0j}^{DL2} + \varepsilon_{ij}^{DL2} \\ ISC_{ij} = \mu_{00}^{ISC} + \mu_{02}^{ISC}SCH_{ij} + \upsilon_{0j}^{ISC} + \varepsilon_{ij}^{ISC} \\ ILS_{ij} = \mu_{00}^{ILS} + \mu_{02}^{ILS}SCH_{ij} + \upsilon_{0j}^{ILS} + \varepsilon_{ij}^{ILS} \\ HWC_{ij} = \mu_{00}^{HWC} + \mu_{02}^{HWC}SCH_{ij} + \upsilon_{0j}^{HWC} + \varepsilon_{ij}^{HWC} \end{cases} \quad (3.6)$$

Now we differentiate the within- and between-variance components of each latent variable in what is equivalent to having random intercepts for factors. Because the variances are directly comparable, the proportion of between-cluster variance to the total variance of a latent variable works as an intraclass correlation for the factor. In addition, since we are interested in comparing latent variables' variances, their metric has to be set by fixing one factor loading of each latent variable to 1. Otherwise, if we fix the latent's variance to 1 and let all loadings be freely estimated, the comparability of the scales is lost (and this incorrect ICC calculation will always yield 0.5 as a result). Indicators still have a between-cluster variance (υ_{0j}). This is their residual variance at Level 2 (Mehta & Neale, 2005).

Why would we do this? First, we end up with a more parsimonious model. Using model fit information, it is possible to compare this more parsimonious model with the one that allows all loadings to be freely estimated and test whether there is a significant difference between the two (Heck & Thomas, 2015). Second, from a theoretical point of view, the between-cluster (school) variance on the latent level (and much less so on the item level) could be an important variable to hypothesize about in a more complex model. We do not present an example of this kind of model with the Dominican PISA survey data, since we have already seen that the factor structure of Level 1 is not reproduced at Level 2, as the former has two dimensions, while the latter is unidimensional. Nevertheless, it is important to understand how this kind of measurement model works because it is used in the next chapter, where we have latent variables with random intercepts as outcomes in structural models.

Multilevel CFA With Random Loadings

So far, we have followed a standard practice in two-level CFA by assuming that the measurement instrument is invariant across groups. However, we may have theoretical reasons to suspect that students from some schools may answer certain questions systematically different from students in other schools. In schools from areas where smartphones are more widespread than computers, students might use their digital devices for browsing the Internet to prepare for classes or for following up on lessons (*ISC* and *ILS*) but would use computers less frequently for homework. Therefore, in those areas, the last variable (*HWC*) would be a bad indicator of the latent construct of use for schoolwork (*SCH*).

One solution, already mentioned, would be multigroup CFA. However, with 186 schools, this would be a very cumbersome model to test, estimate, and interpret results from. A solution, natural for those coming from the MLM perspective, would be to allow factor loadings to also have between-group variance, much like a random slopes regression or path model.

Mathematically, there is nothing to stop one from describing such a model. If we start from Equation 3.1, from earlier in this chapter, now both λ_1 and λ_2 would turn into λ_{1j} and λ_{2j}, decomposed into an average effect and a variance component, denoted, respectively, by μ_{11}, μ_{12}, and υ_{1j}. Equation 3.7 describes this.

$$\begin{cases} \lambda_{1j}^{VID} = \mu_{11}^{VID} + \upsilon_{1j}^{VID} \\[4pt] \lambda_{1j}^{DL1} = \mu_{11}^{DL1} + \upsilon_{1j}^{DL1} \\[4pt] \lambda_{1j}^{DL2} = \mu_{11}^{DL2} + \upsilon_{1j}^{DL2} \\[4pt] \lambda_{2j}^{ISC} = \mu_{12}^{ISC} + \upsilon_{1j}^{ISC} \\[4pt] \lambda_{2j}^{ILS} = \mu_{12}^{ILS} + \upsilon_{1j}^{ILS} \\[4pt] \lambda_{2j}^{HWC} = \mu_{12}^{HWC} + \upsilon_{1j}^{HWC} \end{cases} \quad (3.7)$$

This can be plugged back into Equation 3.3, turning it into the specification presented in Equation 3.8. The reason why, until recently, random loadings models had not been applied more often was the lack of processing power. The traditional maximum likelihood estimators used for CFA and two-level CFA require numeric integration for these models, which may be too demanding computationally. Asparouhov and Muthén (2015) have proposed a Bayesian estimator that makes the process feasible.

Going into the technical details of how Bayesian estimation solves the computational problem is beyond the scope of this introductory text—those properties are described by Asparouhov and Muthén (2015).[8] Instead, we highlight here how random loadings models can be used, as well as a few issues regarding their implementation and interpretation, following B. O. Muthén and Asparouhov (2018).

---

[8] Bayesian estimation for random loadings in two-level measurement models had been incorporated into the item response theory framework earlier for modeling categorical response variables (see De Boeck, 2008; De Jong, Steenkamp, & Fox, 2007; Verhagen & Fox, 2013).

$$\begin{cases} VID_{ij} = \mu_{00}^{VID} + \mu_{11}^{VID}FUNW_{ij} + \mu_{01}^{VID}FUNB_j + v_{1j}^{VID}FUNW_{ij} + \\ \quad + v_{0j}^{VID} + \varepsilon_{ij}^{VID} \\ DL1_{ij} = \mu_{00}^{DL1} + \mu_{11}^{DL1}FUNW_{ij} + \mu_{01}^{DL1}FUNB_j + v_{1j}^{DL1}FUNW_{ij} + \\ \quad + v_{0j}^{DL1} + \varepsilon_{ij}^{DL1} \\ DL2_{ij} = \mu_{00}^{DL2} + \mu_{11}^{DL2}FUNW_{ij} + \mu_{01}^{DL2}FUNB_j + v_{1j}^{DL2}FUNW_{ij} + \\ \quad + v_{0j}^{DL2} + \varepsilon_{ij}^{DL2} \\ ISC_{ij} = \mu_{00}^{ISC} + \mu_{12}^{ISC}SCHW_{ij} + \mu_{02}^{ISC}SCHB_j + v_{1j}^{ISC}SCHW_{ij} + \\ \quad + v_{0j}^{ISC} + \varepsilon_{ij}^{ISC} \\ ILS_{ij} = \mu_{00}^{ILS} + \mu_{12}^{ILS}SCHW_{ij} + \mu_{02}^{ILS}SCHB_j + v_{1j}^{ILS}SCHW_{ij} + \\ \quad + v_{0j}^{ILS} + \varepsilon_{ij}^{ILS} \\ HWC_{ij} = \mu_{00}^{HWC} + \mu_{12}^{HWC}SCHW_{ij} + \mu_{02}^{HWC}SCHB_j + v_{1j}^{HWC}SCHW_{ij} + \\ \quad + v_{0j}^{HWC} + \varepsilon_{ij}^{HWC} \end{cases}$$

(3.8)

Measurement Invariance

Standard two-level CFA with random intercepts assumes measurement invariance in the classical sense of the word. More formally, this means that we have to assume that the within-cluster covariance matrix is the same for all groups. This implies, or is equivalent to, imposing an equality constraint in which the factor loadings for all groups are forced to be the same. By allowing loadings to have a between-group variance component, we relax the invariance assumption. This is equivalent to a test of measurement invariance: If we observe a significant between-group variance of factor loadings, it means that the measurement model is noninvariant across clusters. Therefore, a random loadings model is a quick implementation of a measurement invariance test, very efficient for situations with a large number of groups.

MGCFA invariance testing is done with χ^2 tests of model difference, a very conservative test. Even in the presence of minimal differences across the groups, with additional precision acquired through larger sample sizes, noninvariance quickly becomes significant. A random loadings model is less sensitive to sample sizes and therefore more tolerant of small differences across groups than MGCFA. Conceptually speaking, this approach is less focused on the accumulative differences between every pair of groups and more on overall deviations present across the groups. Moreover, it is implemented the same way regardless of how large the Level 2 sample is and does not require a separate model for each group.

If a measurement model fails the invariance test across groups, parameter estimates (such as means, variances, and regression coefficients) obtained with an equality constraint, or assuming equal loadings across groups, are potentially biased and not to be trusted. Adding random loadings incorporates this variance component, and therefore, it is possible to model and get parameter estimates that are more unbiased without having to resort to the invariance assumption.

The next advantage of a random loadings model is the possibility to explain the between-groups variance of factor loadings. It is possible to add Level 2 covariates that explain the variance of loadings across groups, similar to the random slopes models we cover in Chapters 2 and 4. This expands modeling possibilities to cover more complex measurement and structural relationships with clustered data. For example, we may hypothesize that individuals in more unequal countries understand a certain survey question on redistributive policies differently from those in less unequal countries. A random loadings model allows a researcher to add a measure of inequality (such as the Gini index of income inequality) as a covariate of factor loading variance across groups for any indicator. While adding covariates is done in Chapter 4, the next sections walk through an implementation of a measurement model with random factor loadings.

Example

To present an application of this modeling strategy, we have included random loadings in the model from Figure 3.4. This model had two factors at the within level, with three indicators each, and one at the between level, with six indicators. We add random effects to the within-level factor loadings. The equations that define the new model are presented in Equation 3.9.

$$\begin{cases} VID_{ij} = \mu_{00}^{VID} + \mu_{11}^{VID}FUN_{ij} + \mu_{01}^{VID}DIG_j + \upsilon_{1j}^{VID}FUN_{ij} + \upsilon_{0j}^{VID} + \varepsilon_{ij}^{VID} \\ DL1_{ij} = \mu_{00}^{DL1} + \mu_{11}^{DL1}FUN_{ij} + \mu_{01}^{DL1}DIG_j + \upsilon_{1j}^{DL1}FUN_{ij} + \upsilon_{0j}^{DL1} + \varepsilon_{ij}^{DL1} \\ DL2_{ij} = \mu_{00}^{DL2} + \mu_{11}^{DL2}FUN_{ij} + \mu_{01}^{DL2}DIG_j + \upsilon_{1j}^{DL2}FUN_{ij} + \upsilon_{0j}^{DL2} + \varepsilon_{ij}^{DL2} \\ ISC_{ij} = \mu_{00}^{ISC} + \mu_{12}^{ISC}SCH_{ij} + \mu_{01}^{ISC}DIG_j + \upsilon_{1j}^{ISC}SCH_{ij} + \upsilon_{0j}^{ISC} + \varepsilon_{ij}^{ISC} \\ ILS_{ij} = \mu_{00}^{ILS} + \mu_{12}^{ILS}SCH_{ij} + \mu_{01}^{ILS}DIG_j + \upsilon_{1j}^{ILS}SCH_{ij} + \upsilon_{0j}^{ILS} + \varepsilon_{ij}^{ILS} \\ HWC_{ij} = \mu_{00}^{HWC} + \mu_{12}^{HWC}SCH_{ij} + \mu_{01}^{HWC}DIG_j + \upsilon_{1j}^{HWC}SCH_{ij} + \\ \quad\quad + \upsilon_{0j}^{HWC} + \varepsilon_{ij}^{HWC} \end{cases}$$

(3.9)

Unstandardized results are in Figure 3.5—it is not possible to have standardized results with random slopes or loadings models, only with random intercepts. As we can see, the only parameters estimated at the student level are the indicators' residual variances (ε_{ij}) and the factors covariance. Indicators' intercepts and factor loadings are free to vary across schools, as denoted by the solid black dots in the middle and tip of the arrows.

At the between part, the arrows from the intercept to the λ_{0j} estimates are the overall means for each indicator. For instance, the grand mean of using devices to watch videos (VID) for the whole sample (μ_{00}^{VID}) is 3.262. Estimates below the small curved arrows going from each λ_{0j} into itself are the across-group variances of each indicator intercept (υ_{0j}). The arrows from the intercept to the λ_{1j} and λ_{2j} parameters are the overall average factor loadings of the within-level part (μ_{11} and μ_{12}), while the estimates next to the residuals small curved arrows are the loadings' variances across groups (υ_{1j}). And finally, as before, arrows going from DIG to each λ_{0j} are the between-level factor loadings, or μ_{01}.

We focus the interpretation on the between-level variances of factor loadings. Estimates look fairly small, with the largest being 0.013 for λ_{1j}^{VID}. Even if these are unstandardized estimates, when comparing to the mean factor loading of 1.030 across groups for this indicator, we can interpret it as little cross-cluster variance in loadings. To further this point, we turn to Table 3.2, which contains not only the mean estimate of each loading but also the upper and lower bounds of the 95% Bayesian

Figure 3.5 Two-Level CFA Model of Digital Devices Use With a Unidimensional Level 2 Model and Random Factor Loadings

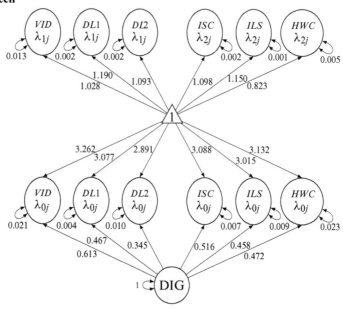

Table 3.2 Random Factor Loadings Estimates for Dominican Republic PISA 2015 Data

Parameter	Mean	95% CI
μ_{11}^{VID}	1.028	[0.985, 1.072]
μ_{11}^{DL1}	1.190	[1.149, 1.225]
μ_{11}^{DL2}	1.093	[1.051, 1.132]
μ_{11}^{ISC}	1.098	[1.055, 1.135]
μ_{11}^{ILS}	1.150	[1.107, 1.190]
μ_{11}^{HWC}	0.823	[0.781, 0.868]

Note: CI = credibility interval.

credibility intervals.[9] It is an indication of how much invariance there is on factor loadings across groups—the smaller the distance between lower and upper bounds, the more invariant the loadings. The range seems small for all indicators. There is none in which, for at least 2.5% of the schools, the indicator worked better or worse than the average by more than a few decimal points.

Model Fit and Comparison

A shortcoming of a Bayesian estimated random loadings CFA is that only one approach to test model fit has been tried and tested to date: the deviance information criterion, or DIC (Spiegelhalter, Best, Carlin, & van der Linde, 2002). Like the AIC and the BIC, the DIC is an incremental indicator of fit, in which lower values indicate better fit. It is sensitive to models with a high number of free parameters and can lead to the selection of less parsimonious models—a shortcoming also of the AIC but less so the BIC. Moreover, the DIC can be used to compare models that are not nested as long as they are on the same sample (Asparouhov & Muthén, 2015).

---

[9] These are conceptually different from frequentist confidence intervals. In Bayesian estimation, we assume that the population parameter is a random variable, and the credibility interval is a 95% range in the estimated posterior distribution of that value.

Table 3.3 DIC Fit Statistics Comparison

	Deviance (DIC)	p_D	Δ_{DIC}
Baseline	57,104.955	346.475	
Random loadings	57,124.916	424.174	19.961

Note: p_D is the estimated effective number of parameters. The random loadings model is presented in Figure 3.5, while its baseline can be found in Figure 3.3.

We can use model fit statistics to test measurement invariance. For a baseline, we run the model from Figure 3.3 with a Bayesian estimator. This allows us to compute the DIC for both models and test if letting loadings vary across groups improves fit. Table 3.3 has model fit information for both. The difference in the DIC indices (Δ_{DIC}) is 19.961. p_D denotes the effective number of parameters and is a model estimate. It is part of the DIC formula, serving as a penalty for model complexity—the more parameters, the higher the DIC. We observe that the DIC of the model that allows loadings to vary across schools is higher than that of the baseline, indicating worse fit. This is caused by the increase in p_D, which is not compensated by a reduced deviance. Therefore, there is measurement invariance in these data, as allowing factor loadings to vary across schools produces a worse-fitting model than constraining them to be the same.

Summary

This chapter provides a fundamental stepping stone for full multilevel structural equation models, by introducing the idea of multilevel measurement models with latent variables. We have seen how clustering can be accounted for in measurement models and how the factor structure might actually differ over different levels of analysis, using the same indicators. And if they do not differ, we explore how factor variance can be separated across the two levels of analysis by keeping the factor metric the same across the levels. Finally, we introduce models with random factor loadings and describe how they increase the modeling potential in multilevel CFA and SEM to account for more complex measurement and theories.

Models tested here have been basic examples, but the technique can be extended seamlessly into more complex ones. For starters, PISA recommends that analyses be done separating public and private schools, something we ignored for simplicity in the examples presented. If one were to take that into account, it is possible to implement a multilevel CFA in a multigroup framework with two groups, one for each kind of school. The interested reader may find more about this in Chapter 5. Furthermore, by switching estimators, one can also apply multilevel CFA to models with categorical indicators. It is possible to add cross-loadings, and the Bayesian approach extends to models with cross-classification (Asparouhov & Muthén, 2015). The most immediate expansion, however, is what we proceed with in Chapter 4: adding covariates, at both levels of analysis, to build multilevel path models with latent variables.

CHAPTER 4. MULTILEVEL STRUCTURAL EQUATION MODELS

Bringing Factor and Path Models Together

Chapters 2 and 3 introduced the multilevel extensions of the main components of structural equation modeling: path analysis and confirmatory factor models. This chapter merges the two into full multilevel structural equation models. We now have a complex causal structure with the potential for mediation and moderation effects, as well as multivariate measurement of constructs of interest. This flexibility allows for the test of our theories while taking into account issues such as measurement error and validity.

To build our examples, we use data from the 2004 Workplace Employment Relations Survey (WERS) teaching data set. It is a survey composed of two questionnaires: one administered to employees and one to company managers. The data we use have interviews with 1,723 managers, one from each company, and a total of 18,918 employees, for an average of 11 employees per firm.[1] The survey is administered in Great Britain and is a random sample from a population composed of all workplaces with five or more employees, including public- and private-sector firms. Alongside cross-national surveys and educational studies, this is another classic multilevel data structure, in which workers are clustered in organizations.[2]

The list with all variables used in this chapter is in Table 4.1. The main outcome is how workers perceive their abilities to be in line with what is expected of them—whether employees consider themselves under- or overqualified for their tasks. We explore the relationship between this outcome and workers' perception of how demanding their job is, how responsive managers of the company are, and their pay. At the company level, we include the number of employees as an exogenous covariate. We must emphasize that the models in this chapter are just an example to demonstrate the method. They are not causally identified and do not intend to make substantive causal claims.

---

[1] We have removed all observations that have missing information on at least one of the variables used in the models. The data set originally contained 22,451 employees and 1,733 managers.

[2] Common extensions include clustering companies into sectors or regions, for example.

Table 4.1 List of Variables and Acronyms From the WERS 2004 Teaching Data

Code	Question	Response Scale
Perception of own skills		
SKL	How well do the work skills you personally have match the skills you need to do your present job?	1 = [my own skills are] much lower to 5 = much higher*
Managerial responsiveness (RES/REB)		
RE1	[Overall, how good would you say managers at this workplace are at] Responding to suggestions from employees or employee representatives	1 = very poor to 5 = very good*
RE2	Allowing employees or employee representatives to influence final decisions	1 = very poor to 5 = very good*
Perception of how hard one's work is (HAR/HAB)		
HW1	My job requires that I work very hard	1 = strongly disagree to 5 = strongly agree*
HW2	I never seem to have enough time to get my work done	1 = strongly disagree to 5 = strongly agree*
HW3	I worry a lot about my work outside working hours	1 = strongly disagree to 5 = strongly agree*
Salary		
PAY	How much do you get paid for your job here, before tax and other deductions are taken out?	1 = £50 or less per week to 14 = £871 or more per week
Workplace characteristic		
NEM	Number of employees in the company (logarithm)	5–9,873 (1.609–9.198)

* indicates variables that were recoded in relation to the original data set so that higher values indicate higher skills, more demanding work, and more responsive management.

Random Intercept of Observed Outcome

Estimating multilevel structural equation models with latent variables follows a similar procedure as that explained in the previous chapters. The total covariance matrix is decomposed into two: within- and between-cluster components, which are additive and uncorrelated, so that $\Sigma_T = \Sigma_W + \Sigma_B$. This has implications for model identification. As discussed in Chapter 1, single-level structural equation models are overidentified if there are more entries in the covariance matrix than the number of free parameters estimated. This remains true in MSEM. However, the heuristic formula to calculate the number of entries in the covariance matrix $p(p + 1)/2$ does not apply anymore. That is because some variables now have only their Level 1 variance estimated, and some can have their variances decomposed into within- and between-level components, and some covariates have variance only at Level 2.

To calculate the number of parameters that can be estimated, we need to calculate the number of entries in each covariance matrix, Σ_W and Σ_B. For both, the formula follows from the one mentioned above, but with one change: In the within case, it becomes $p_W(p_W + 1)/2$, in which p_W is the total number of variables for which we estimate a within-level variance. This includes both those for which *only* a within-level variance is estimated (i.e., we fix its between-level variance to 0) and those that have both within- and between-level variance components estimated.[3] It is similar for the between level, or $p_B(p_B + 1)/2$, in which p_B is the number of variables for which between-level variance is estimated. This includes both contextual variables alone, measured only at the group level, and those measured at the within level but for which we estimate the two variance components. The total number of parameters is given by the formula presented in Equation 4.1.

$$p_W(p_W + 1)/2 + p_B(p_B + 1)/2 + p \quad (4.1)$$

The last p is the total number of variables in the model. It is there because multilevel structural equation models always have a mean structure. The vector of means is always one component of information that counts toward model identification.

---

[3] Every within-level variable that is not group mean centered has at least a small portion of between-group variance. However, as we will see later, we often include variables only in one level of analysis and therefore do not estimate separately their two variance components. We estimate only its total variance and define that as its within variance.

Figure 4.1 Two-Level Model of Perception of Own Skills With a Random Intercept

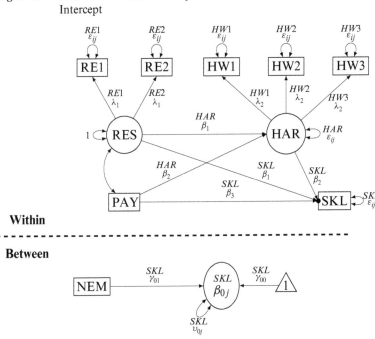

For the examples in this chapter, we want to know what are the determinants of employees feeling that they are under- or overqualified for their jobs. A number of factors can influence this perception. At the individual level, we can think of someone's salary or the nature of activities performed. We can also imagine company-level phenomena influencing perception about one's own skills such as the general work environment or the kind of industry one is in. We therefore have theoretical expectations on both levels of analysis, making it a good example to test with our hierarchical data and a multilevel structural equation model. Our first example is depicted in Figure 4.1. It includes two within-level latent variables: *manager's responsiveness* (RES), measured by two questions answered by employees on how much they think managers take their opinions into account, and *difficulty of work* (HAR), measured by three indicators that ask employees how hard they think their work is and how much they worry about it. The measurement part of this model is described in Equation 4.2.

$$\begin{cases} RE1_{ij} = \overset{RE1}{\lambda_0} + \overset{RE1}{\lambda_1} RES_{ij} + \overset{RE1}{\varepsilon_{ij}} \\ RE2_{ij} = \overset{RE2}{\lambda_0} + \overset{RE2}{\lambda_1} RES_{ij} + \overset{RE2}{\varepsilon_{ij}} \\ HW1_{ij} = \overset{HW1}{\lambda_0} + \overset{HW1}{\lambda_2} HAR_{ij} + \overset{HW1}{\varepsilon_{ij}} \\ HW2_{ij} = \overset{HW2}{\lambda_0} + \overset{HW2}{\lambda_2} HAR_{ij} + \overset{HW2}{\varepsilon_{ij}} \\ HW3_{ij} = \overset{HW3}{\lambda_0} + \overset{HW3}{\lambda_2} HAR_{ij} + \overset{HW3}{\varepsilon_{ij}} \end{cases} \quad (4.2)$$

There is a Level 1 covariate, *PAY*, which is individuals' salaries; the outcome is employees' perceptions of how well their own skills fit their tasks (*SKL*). It indicates whether they think of themselves as over- or underqualified for their job. Their own qualification perception is explained by how hard they think their job is, their salary, and how responsive they perceive the management to be. We also hypothesize that salaries and managers' responsiveness have an impact on how hard individuals think their work is. At the within level, therefore, we have a path model assuming mediated effects. The structural part of this model at Level 1 is given in Equation 4.3.

$$\begin{cases} HAR_{ij} = \overset{HAR}{\beta_0} + \overset{HAR}{\beta_1} RES_{ij} + \overset{HAR}{\beta_2} PAY_{ij} + \overset{HAR}{\varepsilon_{ij}} \\ SKL_{ij} = \overset{SKL}{\beta_{0j}} + \overset{SKL}{\beta_1} RES_{ij} + \overset{SKL}{\beta_2} HAR_{ij} + \overset{SKL}{\beta_3} PAY_{ij} + \overset{SKL}{\varepsilon_{ij}} \end{cases} \quad (4.3)$$

The latent for the perception of how hard work is (*HAR*) has an intercept that is invariant across groups ($\overset{HAR}{\beta_0}$), which, since this is a latent variable for which we must fix the metric, is by default set to 0 and not estimated in the model. HAR_{ij} is explained by the two exogenous variables, one's salary (*PAY*) and the latent for managerial responsiveness (*RES*), with coefficients that are not allowed to vary across workplaces. It also contains a disturbance ($\overset{HAR}{\varepsilon_{ij}}$), which can be freely estimated if we fix one of the factor loadings λ_2 to 1. Skill perception (*SKL*) has an intercept that is allowed to vary across *j* companies ($\overset{SKL}{\beta_{0j}}$), and it is explained by the two latent indicators and by our observed exogenous variable, *PAY*. The degree to which current and needed skills match, *SKL*, also has a residual variance across individuals, $\overset{SKL}{\varepsilon_{ij}}$.

We begin with a simple model for Level 2. Only the intercept of perceived skill is allowed to vary across companies, and it is predicted by a Level 2 covariate: the logged number of employees in a company (*NEM*). This is described in Equation 4.4.

$$\beta_{0j}^{SKL} = \gamma_{00}^{SKL} + \gamma_{01}^{SKL} NEM_j + \upsilon_{0j}^{SKL} \qquad (4.4)$$

Equation 4.4 contains a main intercept, γ_{00}^{SKL}; the coefficient of the Level 2 covariate, γ_{01}^{SKL}; and a between-groups residual variance, υ_{0j}^{SKL}. We can plug this back into the second line of Equation 4.3, which then becomes the specification presented in Equation 4.5.

$$\begin{aligned} SKL_{ij} &= \gamma_{00}^{SKL} + \gamma_{01}^{SKL} NEM_j + \beta_1^{SKL} RES_{ij} + \beta_2^{SKL} HAR_{ij} + \\ &\quad + \beta_3^{SKL} PAY_{ij} + \upsilon_{0j}^{SKL} + \varepsilon_{ij}^{SKL} \end{aligned} \qquad (4.5)$$

Adding the first line of Equation 4.3 to the measurement model in Equation 4.2, along with Equation 4.5, the full set of equations for our first model is given by the specification presented in Equation 4.6.

$$\begin{cases} RE1_{ij} = \lambda_0^{RE1} + \lambda_1^{RE1} RES_{ij} + \varepsilon_{ij}^{RE1} \\[4pt] RE2_{ij} = \lambda_0^{RE2} + \lambda_1^{RE2} RES_{ij} + \varepsilon_{ij}^{RE2} \\[4pt] HW1_{ij} = \lambda_0^{HW1} + \lambda_2^{HW1\,HAR} \beta_1^{HAR} RES_{ij} + \lambda_2^{HW1\,HAR} \beta_2^{HAR} PAY_{ij} + \lambda_2^{HW1\,HAR} \beta_0^{HAR} + \\[2pt] \qquad\quad \lambda_2^{HW1\,HAR} \varepsilon_{ij}^{HAR} + \varepsilon_{ij}^{HW1} \\[4pt] HW2_{ij} = \lambda_0^{HW2} + \lambda_2^{HW2\,HAR} \beta_1^{HAR} RES_{ij} + \lambda_2^{HW2\,HAR} \beta_2^{HAR} PAY_{ij} + \lambda_2^{HW2\,HAR} \beta_0^{HAR} + \\[2pt] \qquad\quad \lambda_2^{HW2\,HAR} \varepsilon_{ij}^{HAR} + \varepsilon_{ij}^{HW2} \\[4pt] HW3_{ij} = \lambda_0^{HW3} + \lambda_2^{HW3\,HAR} \beta_1^{HAR} RES_{ij} + \lambda_2^{HW3\,HAR} \beta_2^{HAR} PAY_{ij} + \lambda_2^{HW3\,HAR} \beta_0^{HAR} + \\[2pt] \qquad\quad \lambda_2^{HW3\,HAR} \varepsilon_{ij}^{HAR} + \varepsilon_{ij}^{HW3} \\[4pt] SKL_{ij} = \gamma_{00}^{SKL} + \gamma_{01}^{SKL} NEM_j + \beta_1^{SKL} RES_{ij} + \beta_2^{SKL} HAR_{ij} + \beta_3^{SKL} PAY_{ij} + \\[2pt] \qquad\quad \upsilon_{0j}^{SKL} + \varepsilon_{ij}^{SKL} \end{cases}$$
$$(4.6)$$

We therefore have six observed endogenous variables, one observed exogenous covariate at Level 1 (PAY), one observed exogenous covariate at Level 2 (NEM), one latent outcome variable (HAR), and one latent exogenous covariate (RES).

Figure 4.2 Two-Level Model of Perception of Own Skills With a Random Intercept

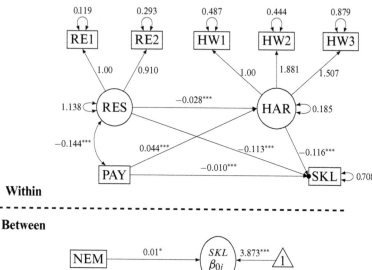

Results in Figure 4.2 show that all Level 1 covariates have a significant negative impact on employees perception of their skills qualification. The more demanding people perceive their job to be, the more they think of themselves as underqualified, which is not surprising. On the flip side, seeing managers as more responsive also explains lower perception of self-skills, and so do higher salaries. The latter might reflect the fact that higher pay is given to those in higher positions within an organization, who have more responsibilities and more demanding tasks. This is suggested by the positive and significant association between one's salary (PAY) and the perception of hard work (HAR). On the other hand, managers' responsiveness is negatively associated with job difficulty. At the between level, companies where employees feel more overqualified are, on average, larger.

Moreover, the model shows good fit statistics. The χ^2 test is significant, which is not surprising given the more than 18,000 observations in our sample. On the other hand, RMSEA is 0.046, and CFI and TLI are, respectively, 0.985 and 0.968. SRMR, which in two-level models gives separate values for within and between components, is 0.028 for the within part and 0.008 for the between level.

Multilevel Latent Covariate Model

In standard multilevel modeling, it is common to find situations in which a Level 1 variable, centered on its group means, serves as a covariate at Level 1, and its group means are entered into the model as a Level 2 covariate. For example, employee reports of workplace climate vary at Level 1 but can be aggregated also at the company level to indicate an average climate for each firm. These are called *contextual variables* (Boyd & Iversen, 1979; Raudenbush & Bryk, 2002) but are sometimes encountered as *configural constructs* as well (Kozlowski & Klein, 2000).

However, the aggregation of Level 1 characteristics using only group means assumes no measurement error. This can be problematic, especially given the relatively small average cluster size. It is likely that the perception of a few employees, on average, might not be representative of overall climate in a company with hundreds of workers.

As we point out in the introduction, one of the great advantages of structural equation modeling is its incorporation of measurement error into causal models. Merging this framework with multilevel models is called the *multilevel latent covariate model* (Lüdtke et al., 2008; Marsh et al., 2009). In these, instead of simply taking the group mean and using it as the Level 2 contextual covariate, we decompose the Level 1 covariate into within- and between-variance components, along with its overall intercept (or grand mean, in a throwback to the multilevel modeling literature).

In our current model, we can imagine not only that employees' salaries affect their perceived skills but also that the average salary of those around them may affect skill perception. The average company salary, therefore, would be a company characteristic, a contextual covariate formed by the aggregation of employees' individual pay. Applying the variance decomposition described in the previous paragraph to our Level 1 observed covariate PAY results in the specification presented in Equation 4.7:

$$PAY_{ij} = \gamma_{00}^{PAY} + \upsilon_{0j}^{PAY} + \varepsilon_{ij}^{PAY} \qquad (4.7)$$

The decomposition is done the same way as it was with the outcome variable in a random intercepts model, except that here we refer to the *mean* of pay instead of its intercept since it is still exogenous. We aggregate it as a latent contextual covariate with a mean and a variance across groups, which is estimated as a model parameter, υ_{0j}^{PAY}.

A classic application of such models is in the educational literature, in findings referred to as 'Big Fish Little Pond Effects' (BFLPE; Marsh & Parker, 1984). In these models, students' abilities predict their self-concept, and the average ability in a classroom (aggregated from students) is used to explain the average self-concept in classes. The effect refers to the observation that, while students with high abilities have a better self-concept, their self-concept is worse in schools or classrooms full of high achievers than in environments with low average student ability (making the good student in a low-achieving school the proverbial big fish in the little pond). Latent contextual covariate models were first developed exactly with these applications in mind (Lüdtke et al., 2008; Marsh et al., 2009).

We may expect a similar scenario in the organizational context. In the previous section, we found that higher salaries predict lower perception of self-skills. There might be two reasons for this. First, it may happen because people with higher salaries have more demanding tasks, which makes it an individual-level phenomenon. Or, it is because people in companies that pay better, and therefore are expected to deliver more complex services having more qualified personnel overall, see themselves as less skilled in comparison to their also highly qualified coworkers. This would make it a company-level effect. The model we present with results in Figure 4.3 is used to test which of these explanations apply. In it, we regress perceived skill (SKL) on salary (PAY) at the within level testing the first hypothesis, and at the between level, we regress the across-companies variance of people's perception of their own skill, or β_{0j}^{SKL}, on the across-companies variance of people's salaries, which we denote by β_{0j}^{PAY}. This tests the second hypothesis: whether higher average salaries lead to lower average self-perceived skills. Equation 4.8 presents the structural part of the Level 2 model.

$$\beta_{0j}^{SKL} = \gamma_{00}^{SKL} + \gamma_{01}^{SKL} NEM_j + \gamma_{02}^{SKL} \beta_{0j}^{PAY} + \upsilon_{0j}^{SKL} \qquad (4.8)$$

This formula adds the term $\gamma_{02}^{SKL} \beta_{0j}^{PAY}$ to Equation 4.4. We can also add it to the last line of Equation 4.5, which defined SKL_{ij} for the previous model, to obtain the specification presented in Equation 4.9:

$$\begin{aligned} SKL_{ij} &= \gamma_{00}^{SKL} + \gamma_{01}^{SKL} NEM_j + \gamma_{02}^{SKL} \beta_{0j}^{PAY} + \beta_1^{SKL} RES_{ij} + \beta_2^{SKL} HAR_{ij} \\ &+ \beta_3^{SKL} PAY_{ij} + \upsilon_{0j}^{SKL} + \varepsilon_{ij}^{SKL} \end{aligned} \qquad (4.9)$$

In the lower part of the figure, we see that β_{0j}^{PAY} has a mean (γ_{00}^{PAY}) of 8.197 and a variance ($\upsilon_{0j}^{PAY} = 4.878$). At the within part, PAY has a within-variance component ε_{ij}^{PAY} that equals 6.195. Back at the between

Figure 4.3 Latent Contextual Covariate Model

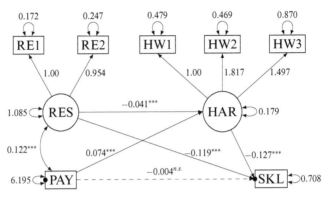

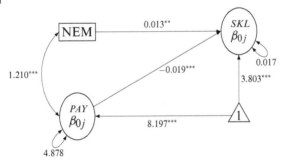

part, it also explains the variation in the random intercept of skill with a coefficient γ_{02}^{SKL}.

This model has good fit, with slightly better fit indicators than the one that did not estimate the between-level variance of *PAY* in the previous section. RMSEA is 0.044; CFI and TLI are 0.988 and 0.972, respectively; and SRMR is 0.026 for the within level and again 0.008 for the between level.[4] Looking at the results, almost all coefficients and estimates remain the same compared to the previous model, which had no contextual covariate. One important difference, however, is observed: The relationship between *PAY* and perceived skill (*SKL*) is not significant at the within level (within the firms), and that coefficient is very small

---

[4] An assessment of comparative fit indices is done at the end of this chapter.

(−.004), less than half of what it was before. The relationship at the between level, however, is significant. This indicates that higher average salaries in a company make people working there feel less qualified for their jobs. Higher average salaries might be indicative of companies that deliver more complex services or products and in which one is surrounded by well-qualified professionals. This, in turn, might make individuals perceive themselves as less skilled. Or, conversely, individuals in companies with lower average salaries, which deliver more mundane services, may feel more qualified for their tasks.

To test whether a relationship between two variables is stronger at the Level 1 or Level 2, one can model the difference between the within- and between-level coefficients ($\beta_3^{SKL} - \gamma_{02}^{SKL}$) and estimate this simultaneously with the rest of the model, obtaining confidence intervals and a significance test. In this case, the difference of .015 is statistically significant, meaning that most of the relationship between *PAY* and one's perception of his or her own skill (*SKL*) is at the company, and not employee, level. For the first model, in Figure 4.2, we interpreted the negative within-level relationship between *PAY* and skill (*SKL*) as individuals in higher positions, with more responsibilities and complex tasks, feeling they are not up to it. These findings, however, show that what matters more is the average pay of those around the employee and the nature of the company itself, rather than of one's individual position.

Structural Models With Between-Level Latent Variables

Salaries are not the only factor that may have an influence both at the individual level and as a company characteristic. Perception of managerial responsiveness and work difficulty can also be translated into company attributes: Some companies are more demanding than others, and some have a more responsive managerial culture than others. For example, a researcher might be interested in whether self-perceived skills are more dependent on people's own perceptions of their managers' responsiveness or on the company in general having a more responsive managerial style. The only difference from the example in the previous section separating the within- and between-level impact of salaries is that managerial responsiveness and work difficulty are measured at the individual level as latent variables.

To separate latent variables between the two levels of analysis in a meaningful way, we now also form comparable latent variables at the company level using the intercepts of the indicators measured at the individual level. This model is depicted in Figure 4.4. *REB* is interpreted as a culture of responsiveness by managers in companies. *HAB* is interpreted as how much a company is demanding from its workers. What we end up with is the firm-level equivalent of individuals' perception of managerial responsiveness (our old *RES* variable) and the individual perception of how hard one's work is (or *HAR*). Basically, once we aggregate up the individual indicator intercepts to the firm level, they can serve as factor indicators and can produce a latent for the firm-level characteristic.

At this stage, we do not yet assume cross-level invariance, so there are no equality constraints on factor loadings across Level 1 and Level 2 indicators (as described in Chapter 3). The imposition of these equality constraints will be done in the next section.

For now, at the firm level, the indicators of *REB* and *HAB* are the intercepts (λ_{0j}) for each within-level (individual level) indicator, just as we have seen in the previous chapter on multilevel measurement models. They can be interpreted as company characteristics, but they are not necessarily equal to an aggregation of individual characteristics. The measurement part of this model is defined in Equation 4.10:

$$\begin{cases} \lambda_{0j}^{RE1} = \mu_{00}^{RE1} + \mu_{01}^{RE1} REB_j + \upsilon_{0j}^{RE1} \\[4pt] \lambda_{0j}^{RE2} = \mu_{00}^{RE2} + \mu_{01}^{RE2} REB_j + \upsilon_{0j}^{RE2} \\[4pt] \lambda_{0j}^{HW1} = \mu_{00}^{HW1} + \mu_{02}^{HW1} HAB_j + \upsilon_{0j}^{HW1} \\[4pt] \lambda_{0j}^{HW2} = \mu_{00}^{HW2} + \mu_{02}^{HW2} HAB_j + \upsilon_{0j}^{HW2} \\[4pt] \lambda_{0j}^{HW3} = \mu_{00}^{HW3} + \mu_{02}^{HW3} HAB_j + \upsilon_{0j}^{HW3} \end{cases} \quad (4.10)$$

Our main substantive goal is to investigate how managerial responsiveness and work difficulty, as company characteristics, influence workers' average skill perceptions. For this reason, we also add the structural relationships to the between part of this model. We maintain the random intercept for perceived skill (*SKL*) from the previous model, but now its between-companies variance is explained by the two between-level latent variables, besides the between variance of salaries (υ_{0j}^{PAY}) and the Level 2 covariate: the number of employees. In addition, we hypothesize that the number of employees should be associated with a company's responsiveness—larger companies might naturally have more difficulty consulting and responding to demands from individual

Figure 4.4 Random Intercepts Model With Between-Level Latent Variables

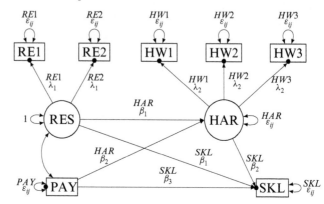

Within

- -

Between

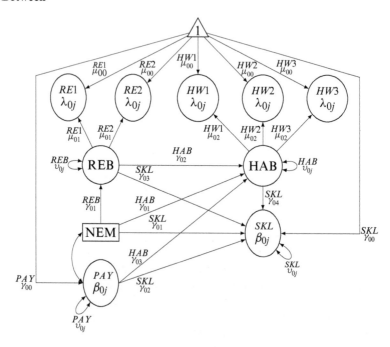

workers. Furthermore, we hypothesize that at the between level as well, responsiveness is negatively related to how demanding the job is and affects workers' perception of their own skills. The number of employees is also expected to have an effect on how demanding a company is. This set of between-level structural relationships updates Equation 4.8, for β_{0j}^{SKL}, adding now the two new endogenous latent variables for the between level, HAB_j and REB_j, with their respective covariates.[5]

$$\begin{cases} \beta_{0j}^{SKL} = \gamma_{00}^{SKL} + \gamma_{01}^{SKL} NEM_j + \gamma_{02}^{SKL} \beta_{0j}^{PAY} + \gamma_{03}^{SKL} REB_j + \gamma_{04}^{SKL} HAB_j + \upsilon_{0j}^{SKL} \\ HAB_j = \gamma_{00}^{HAB} + \gamma_{01}^{HAB} NEM_j + \gamma_{02}^{HAB} REB_j + \gamma_{03}^{HAB} \beta_{0j}^{PAY} + \upsilon_{0j}^{HAB} \\ REB_j = \gamma_{00}^{REB} + \gamma_{01}^{REB} NEM_j + \upsilon_{0j}^{REB} \end{cases}$$
(4.11)

If we combine this structural part (Equation 4.11) into the Level 2 measurement model (Equation 4.10), we get the specification presented in Equation 4.12.

$$\begin{cases} \lambda_{0j}^{RE1} = \mu_{00}^{RE1} + \mu_{01}^{RE1} \gamma_{00}^{REB} + \mu_{01}^{RE1} \gamma_{01}^{REB} NEM_j + \mu_{01}^{RE1} \upsilon_{0j}^{REB} + \upsilon_{0j}^{RE1} \\ \lambda_{0j}^{RE2} = \mu_{00}^{RE2} + \mu_{01}^{RE2} \gamma_{00}^{REB} + \mu_{01}^{RE2} \gamma_{01}^{REB} NEM_j + \mu_{01}^{RE2} \upsilon_{0j}^{REB} + \upsilon_{0j}^{RE2} \\ \lambda_{0j}^{HW1} = \mu_{00}^{HW1} + \mu_{02}^{HW1} \gamma_{00}^{HAB} + \mu_{02}^{HW1} \gamma_{01}^{HAB} NEM_j + \mu_{02}^{HW1} \gamma_{02}^{HAB} REB_j + \\ \qquad \mu_{02}^{HW1} \upsilon_{0j}^{HAB} + \upsilon_{0j}^{HW1} \\ \lambda_{0j}^{HW2} = \mu_{00}^{HW2} + \mu_{02}^{HW2} \gamma_{00}^{HAB} + \mu_{02}^{HW2} \gamma_{01}^{HAB} NEM_j + \mu_{02}^{HW2} \gamma_{02}^{HAB} REB_j + \\ \qquad \mu_{02}^{HW2} \upsilon_{0j}^{HAB} + \upsilon_{0j}^{HW2} \\ \lambda_{0j}^{HW3} = \mu_{00}^{HW3} + \mu_{02}^{HW3} \gamma_{00}^{HAB} + \mu_{02}^{HW3} \gamma_{01}^{HAB} NEM_j + \mu_{02}^{HW3} \gamma_{02}^{HAB} REB_j + \\ \qquad \mu_{02}^{HW3} \upsilon_{0j}^{HAB} + \upsilon_{0j}^{HW3} \end{cases}$$
(4.12)

---

[5] One must note that, even though we add an intercept to the between part of Figure 4.4, there are no arrows going into the two latent variables REB and HAB, which in Equation 4.11 would be the parameters γ_{00}^{REB} and γ_{00}^{HAB}. That is because, when setting the metric for these latent variables, the overall mean is fixed to 0 and not estimated in the model. We have the parameters in the equations only for completeness.

Here, each indicator's group-varying intercept (say, λ_{0j}^{RE1}) is formed by its overall mean (μ_{00}^{RE1}) and the multiplication of the factor loading (μ_{01}^{RE1}) by the elements that make up the total variance of the between-level latent variable: for the first line, REB_j. They are on the right-hand side of Equation 4.11. There, we multiply μ_{01}^{RE1} by ($\gamma_{00}^{REB} + \gamma_{01}^{REB} NEM_j + \upsilon_{0j}^{REB}$). Adding, finally, the between-level residual variance of $RE1$, υ_{0j}^{RE1}, we have the first line in Equation 4.12.

Furthermore, on the last three rows of Equation 4.12, we have left REB_j, in that case as the covariate for HAB, which is the latent variable formed by $\lambda_{0j}^{HW1} - \lambda_{0j}^{HW3}$. At the same time, we have already specified a statistical model for REB_j, spelled out in the last line of Equation 4.11. If we replace this model for REB_j in the last three lines of Equation 4.12, these last three lines then appear as presented in Equation 4.13.

$$\begin{cases} \lambda_{0j}^{HW1} = \mu_{00}^{HW1} + \mu_{02}^{HW1\ HAB} \gamma_{00}^{HAB} + \mu_{02}^{HW1\ HAB} \gamma_{01}^{HAB} NEM_j + \mu_{02}^{HW1\ HAB} \gamma_{02}^{REB} \gamma_{00}^{REB} + \mu_{02}^{HW1\ HAB} \gamma_{03}^{PAY} \beta_{0j} + \\ \quad \mu_{02}^{HW1\ HAB} \gamma_{02}^{REB} \gamma_{01}^{REB} NEM_j + \mu_{02}^{HW1\ HAB} \gamma_{02}^{REB} \upsilon_{0j}^{REB} + \mu_{02}^{HW1\ HAB} \upsilon_{0j} + \upsilon_{0j}^{HW1} \\[6pt] \lambda_{0j}^{HW2} = \mu_{00}^{HW2} + \mu_{02}^{HW2\ HAB} \gamma_{00}^{HAB} + \mu_{02}^{HW2\ HAB} \gamma_{01}^{HAB} NEM_j + \mu_{02}^{HW2\ HAB} \gamma_{02}^{REB} \gamma_{00}^{REB} + \mu_{02}^{HW2\ HAB} \gamma_{03}^{PAY} \beta_{0j} + \\ \quad \mu_{02}^{HW2\ HAB} \gamma_{02}^{REB} \gamma_{01}^{REB} NEM_j + \mu_{02}^{HW2\ HAB} \gamma_{02}^{REB} \upsilon_{0j}^{REB} + \mu_{02}^{HW2\ HAB} \upsilon_{0j} + \upsilon_{0j}^{HW2} \\[6pt] \lambda_{0j}^{HW3} = \mu_{00}^{HW3} + \mu_{02}^{HW3\ HAB} \gamma_{00}^{HAB} + \mu_{02}^{HW3\ HAB} \gamma_{01}^{HAB} NEM_j + \mu_{02}^{HW3\ HAB} \gamma_{02}^{REB} \gamma_{00}^{REB} + \mu_{02}^{HW3\ HAB} \gamma_{03}^{PAY} \beta_{0j} + \\ \quad \mu_{02}^{HW3\ HAB} \gamma_{02}^{REB} \gamma_{01}^{REB} NEM_j + \mu_{02}^{HW3\ HAB} \gamma_{02}^{REB} \upsilon_{0j}^{REB} + \mu_{02}^{HW3\ HAB} \upsilon_{0j} + \upsilon_{0j}^{HW3} \end{cases}$$

(4.13)

As the left-hand side of these equations indicates, they are only the between-group variance in indicators' intercepts. Therefore, to complete the model, we can replace the intercepts of our indicators in Equation 4.6 (λ_0) by their new formulations, the right-hand sides of Equation 4.12 for λ_{0j}^{RE1} and λ_{0j}^{RE2}, and the right-hand sides of Equation 4.13 for λ_{0j}^{HW1}, λ_{0j}^{HW2}, and λ_{0j}^{HW3}. The full model for Figure 4.4, where all entries in the left-hand side are observed variables, can be written in equation format (although, at this point, only with an increased appreciation for the parsimony associated with the matrix algebraic formulation), as shown in Equation 4.14.

$$\begin{cases}
RE1_{ij} = \mu_{00}^{RE1} + \lambda_1^{RE1} RES_{ij} + \mu_{01}^{RE1} \gamma_{00}^{REB} + \mu_{01}^{RE1} \gamma_{01}^{REB} NEM_j + \mu_{01}^{RE1} \upsilon_{0j}^{REB} + \upsilon_{0j}^{RE1} + \varepsilon_{ij}^{RE1} \\
RE2_{ij} = \mu_{00}^{RE2} + \lambda_1^{RE2} RES_{ij} + \mu_{01}^{RE2} \gamma_{00}^{REB} + \mu_{01}^{RE2} \gamma_{01}^{REB} NEM_j + \mu_{01}^{RE2} \upsilon_{0j}^{REB} + \upsilon_{0j}^{RE2} + \varepsilon_{ij}^{RE2} \\
HW1_{ij} = \mu_{00}^{HW1} + \mu_{02}^{HW1\,HAB} \gamma_{00}^{HW1\,HAB} + \mu_{02}^{HW1\,HAB} \gamma_{01}^{HW1\,HAB} NEM_j + \mu_{02}^{HW1\,HAB} \gamma_{02}^{HW1\,HAB\,REB} \gamma_{00}^{HW1\,HAR} + \mu_{02}^{HW1\,HAB} \gamma_{02}^{HW1\,HAB\,REB} \gamma_{01}^{HW1\,HAR} NEM_j + \mu_{02}^{HW1\,HAB} \gamma_{02}^{HW1\,HAB\,REB} \upsilon_{0j}^{HW1\,HAR} + \\
\quad \mu_{02}^{HW1\,HAB\,PAY} \beta_{0j}^{HW1\,HAR} + \mu_{02}^{HW1\,HAB} \upsilon_{0j}^{HW1} + \lambda_2^{HW1\,HAR} \beta_0 + \lambda_2^{HW1\,HAR} \beta_1 RES_{ij} + \lambda_2^{HW1\,HAR} \beta_2 PAY_{ij} + \lambda_2^{HW1\,HAR} \varepsilon_{ij} + \upsilon_{0j}^{HW1} + \varepsilon_{ij}^{HW1} \\
HW2_{ij} = \mu_{00}^{HW2} + \mu_{02}^{HW2\,HAB} \gamma_{00}^{HW2\,HAB} + \mu_{02}^{HW2\,HAB} \gamma_{01}^{HW2\,HAB} NEM_j + \mu_{02}^{HW2\,HAB} \gamma_{02}^{HW2\,HAB\,REB} \gamma_{00}^{HW2\,HAR} + \mu_{02}^{HW2\,HAB} \gamma_{02}^{HW2\,HAB\,REB} \gamma_{01}^{HW2\,HAR} NEM_j + \mu_{02}^{HW2\,HAB} \gamma_{02}^{HW2\,HAB\,REB} \upsilon_{0j}^{HW2\,HAR} + \\
\quad \mu_{02}^{HW2\,HAB\,PAY} \beta_{0j}^{HW2\,HAR} + \mu_{02}^{HW2\,HAB} \upsilon_{0j}^{HW2} + \lambda_2^{HW2\,HAR} \beta_0 + \lambda_2^{HW2\,HAR} \beta_1 RES_{ij} + \lambda_2^{HW2\,HAR} \beta_2 PAY_{ij} + \lambda_2^{HW2\,HAR} \varepsilon_{ij} + \upsilon_{0j}^{HW2} + \varepsilon_{ij}^{HW2} \\
HW3_{ij} = \mu_{00}^{HW3} + \mu_{02}^{HW3\,HAB} \gamma_{00}^{HW3\,HAB} + \mu_{02}^{HW3\,HAB} \gamma_{01}^{HW3\,HAB} NEM_j + \mu_{02}^{HW3\,HAB} \gamma_{02}^{HW3\,HAB\,REB} \gamma_{00}^{HW3\,HAR} + \mu_{02}^{HW3\,HAB} \gamma_{02}^{HW3\,HAB\,REB} \gamma_{01}^{HW3\,HAR} NEM_j + \mu_{02}^{HW3\,HAB} \gamma_{02}^{HW3\,HAB\,REB} \upsilon_{0j}^{HW3\,HAR} + \\
\quad \mu_{02}^{HW3\,HAB\,PAY} \beta_{0j}^{HW3\,HAR} + \mu_{02}^{HW3\,HAB} \upsilon_{0j}^{HW3} + \lambda_2^{HW3\,HAR} \beta_0 + \lambda_2^{HW3\,HAR} \beta_1 RES_{ij} + \lambda_2^{HW3\,HAR} \beta_2 PAY_{ij} + \lambda_2^{HW3\,HAR} \varepsilon_{ij} + \upsilon_{0j}^{HW3} + \varepsilon_{ij}^{HW3} \\
SKL_{ij} = \gamma_{00}^{SKL} + \gamma_{01}^{SKL} NEM_j + \beta_1^{SKL} RES_{ij} + \beta_2^{SKL\,HAR} \beta_0 + \beta_2^{SKL\,HAR} \beta_1 RES_{ij} + \beta_2^{SKL\,HAR} \beta_2 PAY_{ij} + \beta_2^{SKL\,HAR} \varepsilon_{ij} + \\
\quad \beta_3^{SKL} PAY_{ij} + \gamma_{02}^{SKL\,PAY} \beta_{0j}^{SKL\,REB} + \gamma_{03}^{SKL\,REB} \gamma_{00}^{SKL\,REB} + \gamma_{03}^{SKL\,REB} \gamma_{01}^{SKL\,REB} NEM_j + \gamma_{03}^{SKL\,REB} \upsilon_{0j}^{SKL\,REB} + \gamma_{04}^{SKL\,HAB} \gamma_{00}^{SKL\,HAB} + \gamma_{04}^{SKL\,HAB} \gamma_{01}^{SKL\,HAB} NEM_j + \\
\quad \gamma_{04}^{SKL\,HAB\,REB} \gamma_{02}^{SKL\,HAB\,REB} + \gamma_{04}^{SKL\,HAB\,REB} \gamma_{02}^{SKL\,HAB\,REB} \gamma_{01}^{SKL} NEM_j + \gamma_{04}^{SKL\,HAB} \upsilon_{0j}^{SKL\,HAB} + \upsilon_{0j}^{SKL} + \varepsilon_{ij}^{SKL}
\end{cases}$$

(4.14)

Note that, in this, we have also replaced HAB_j and REB_j by their defining equations (Equation 4.11) for SKL_{ij} and β_{0j}^{SKL}. For example, a term like $\gamma_{03}^{SKL} REB_j$ is replaced by $\gamma_{03}^{SKL} * (\gamma_{00}^{REB} + \gamma_{01}^{REB} NEM_j + \upsilon_{0j}^{REB})$. At the end, only exogenous variables and estimated model parameters are left on the right-hand side of Equation 4.14.

Results in Figure 4.5 show that the structure of relationships at the within level does not change much compared to the model in the previous section. Model fit is also very similar, with the one exception of SRMR-between, which goes up to 0.072. In this model, we observe several significant associations at the between level. As theorized, companies with more employees are less responsive. On the other hand, company size is unrelated to how demanding companies are. And, at the company level, managerial responsiveness is unrelated to how demanding a company is. Furthermore, both latent variables, which have significant negative effects on SKL at Level 1, also have significant and negative effects on β_{0j}^{SKL} at Level 2. However, it is important to notice that, in this case, we cannot directly subtract the between-level coefficient from the within, as done in the previous model. That is because the latent variables in each level are on a different metric—we have not constrained factor loadings to be the same at the two levels. That is done in the next section.

Random Intercepts of Latent Variables

To directly compare the size of coefficients between the two levels, we must assume cross-level measurement invariance (Marsh et al., 2009), using the set of equality constraints of factor loadings discussed in Chapter 3:

$$\begin{cases} \lambda_1^{RE1} = \mu_{01}^{RE1} \\[4pt] \lambda_1^{RE2} = \mu_{01}^{RE2} \\[4pt] \lambda_2^{HW1} = \mu_{02}^{HW1} \\[4pt] \lambda_2^{HW2} = \mu_{02}^{HW2} \\[4pt] \lambda_2^{HW3} = \mu_{02}^{HW3} \end{cases} \quad (4.15)$$

95

Figure 4.5 Random Intercepts Model With Between-Level Latent Variables (unstandardized results)

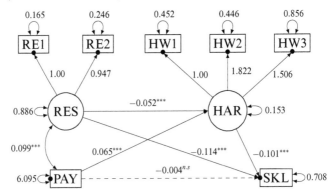

Within

Between

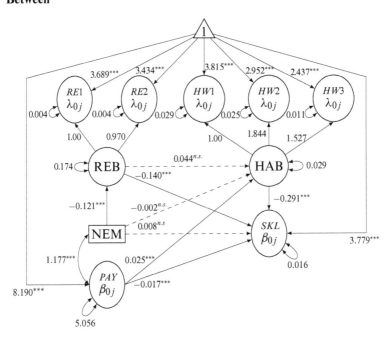

This way, the latent variables on both levels have the same metric, and the sizes of the relationships between them and other variables can be directly summed. It becomes possible, therefore, to test whether the relationship between managers responsiveness and job difficulty, for example, is mostly a worker-level or a company-level characteristic.

In Figure 4.6, random latent variable means at the within level are denoted with solid dots inside the circles. Once again, their overall mean is fixed to 0, and so no arrow goes from the intercept (triangle) to the latent variables. The equality constraints introduced by Equation 4.15 mean that the latent variables at the within and between level have the same metric, and therefore, their variances can be directly compared and added. It is effectively splitting the total variance of, say, RES into RES_W and RES_B, allowing us to model the variation in latent variable intercepts and to compare regression coefficients across levels to test whether effects are mainly individual or contextual. This is an example of a so-called *doubly latent model* (Marsh et al., 2009).

Results in this model are similar to those observed in the previous one. This is not unexpected as factor loadings at the within and between levels were already similar, so that an equality constraint does not change the model too much. Fit indicators remain virtually the same, confirming the two models are very similar. However, we can now perform tests comparing within- and between-level coefficients to investigate whether the relationships between our variables are mostly individual or contextual.

As in the model tested previously in Figure 4.3, we can estimate whether a difference between Level 1 and Level 2 coefficients is statistically significant. This tells us whether a given relationship exists at the employee or at the company level, or if it is perhaps roughly evenly spread across the two. The coefficients of SKL on RES and REB are not significantly different from the within to the between levels (at $-.119$ and $-.139$, respectively). This means that the significant effect of higher responsiveness in relation to lower self-perceived skills is evenly spread at both the individual and the contextual levels. There is a significant difference, however, in the effect between the two latent variables. At the individual level, having more responsive managers leads workers to think their job is less demanding ($\beta_1^{HAR} = -0.052^*$). At the company level, this is a positive and significant relationship ($\gamma_{02}^{HAR} = 0.045^*$). The difference between these two coefficients is significant (-0.099^*). It indicates that managerial responsiveness leads to workers perceiving their jobs to be easier but that companies with more responsive managers are those in which work is more demanding.

Figure 4.6 Random Intercepts Model With Random Mean Latent Variables

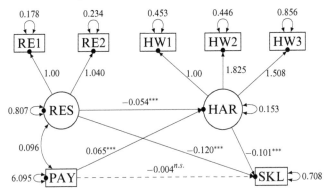

Within

- -

Between

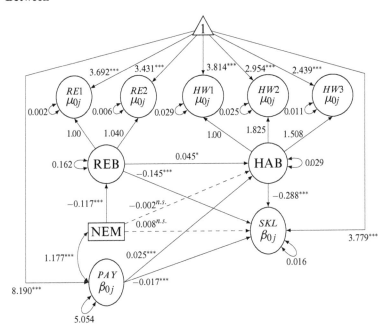

This appears to be a case of Simpson's paradox: The relationship between two variables is inverted, depending on the level of analysis one looks at. We may speculate about the mechanism at play the following way. Managerial responsiveness makes tasks seem less complicated for employees. It is easier to perform your duties when supervisors are more attentive to your questions or concerns. Therefore, at the individual level, the more people perceive their managers as responsive, the less difficult they think their job is. Companies that deliver services demanding complex work know that fact and therefore foster a culture of responsiveness to better achieve their results. Hence, on average, companies with more demanding tasks will also have higher levels of responsiveness, even if within the companies higher perceived responsiveness leads to perceptions of easier work.

There are two additional relationships that are modeled on both levels and have a stronger Level 2 component than Level 1. These are the impact of PAY and perceived work difficulty (HAR/HAB) on perceived skill (SKL). The company-level effect of PAY has been discussed previously. The latter suggests that workers feeling more underqualified are driven more by them working at more demanding companies than by their personal sense of their own job being difficult.

Random Slopes MSEM

As the reader has seen, some of our relationships have distinct within- and between-level components, but they may also vary from company to company. In the next model, we hypothesize that the effect of people's salary on how hard they think their job is varies across companies. In some, high salary differences between employees may accompany highly different levels of work complexity, which might not be the case in other firms. We move, therefore, to the model depicted in Figure 4.7. This has a few differences compared to the previous model, from Figure 4.6. First of all, the random slope between PAY and how hard people find work (HAR) is denoted, at the within level, with the solid dot in the middle of the arrow. This estimate enters the between part of the model as a latent variable, denoted by a circle, β_{2j}^{HAR}. It has an overall intercept, or an overall average effect of people's PAY on the perception of work difficulty (HAR) across all companies (γ_{20}^{HAR}), and a between-level residual variance v_{2j}^{HAR}. The variance in this slope is explained by the number of employees in a company.

Figure 4.7 Random Slopes Model With Random Mean Latent Variables

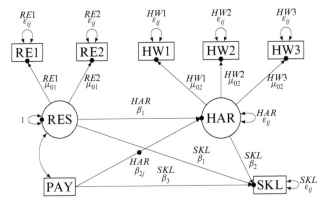

Within

- -

Between

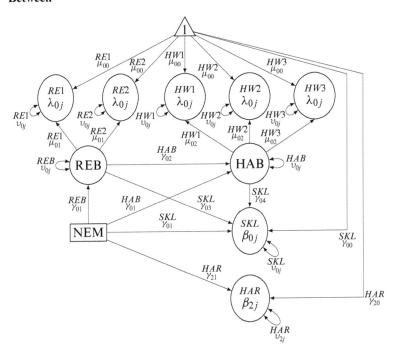

Formally, this difference indicates that β_{2j}^{HAR} can be written up as seen in Equation 4.16.

$$\beta_{2j}^{HAR} = \gamma_{20}^{HAR} + \gamma_{21}^{HAR} NEM_j + \upsilon_{2j}^{HAR} \qquad (4.16)$$

This formulation transforms the overall specification that defined SKL into what can be seen in Equation 4.17.

$$\begin{aligned}
SKL_{ij} =\ & \gamma_{00}^{SKL} + \gamma_{01}^{SKL} NEM_j + \beta_1^{SKL} RES_{ij} + \beta_2^{SKL} \beta_0^{HAR} + \\
+\ & \beta_2^{SKL} \beta_1^{HAR} RES_{ij} + \beta_2^{SKL} \gamma_{20}^{HAR} PAY_{ij} + \\
+\ & \beta_2^{SKL} \gamma_{21}^{HAR} PAY_{ij} NEM_j + \beta_2^{SKL} \upsilon_{2j}^{HAR} PAY_{ij} + \beta_2^{SKL} \varepsilon_{ij}^{HAR} + \\
+\ & \beta_3^{SKL} PAY_{ij} + \gamma_{03}^{SKL\ REB} \gamma_{00}^{REB} + \gamma_{03}^{SKL\ REB} \gamma_{01}^{REB} NEM_j + \gamma_{03}^{SKL\ REB} \upsilon_{0j}^{REB} + \\
+\ & \gamma_{04}^{SKL\ HAB} \gamma_{00}^{HAB} + \gamma_{04}^{SKL\ HAB} \gamma_{01}^{HAB} NEM_j + \gamma_{04}^{SKL\ HAB\ REB} \gamma_{02}^{HAB} \gamma_{00}^{REB} + \\
+\ & \gamma_{04}^{SKL\ HAB\ REB} \gamma_{02}^{HAB} \gamma_{01}^{REB} NEM_j + \gamma_{04}^{SKL\ HAB\ REB} \gamma_{02}^{HAB} \upsilon_{0j}^{REB} + \gamma_{04}^{SKL\ HAB} \upsilon_{0j}^{HAB} + \\
+\ & \upsilon_{0j}^{SKL} + \varepsilon_{ij}^{SKL} \qquad (4.17)
\end{aligned}$$

In the meantime, the equations defining the indicators for how hard one's work is (HAR) are also changed. They previously contained β_2^{HAR} multiplying PAY, which was used to explain HAR. Substituting the entire formula from Equation 4.16 into the place of β_{2j}^{HAR} in Equation 4.14, they are now defined as in Equation 4.18. Another important difference from the model in Figure 4.6 is that we do not allow the between-level variance of PAY to be estimated. This means it is no longer a latent covariate.[6]

Results are in Figure 4.8. Since we removed the between-level component of the relationship between payment and skills perception, the coefficient is once again significant at the employee level, as it was in the first model, in Figure 4.2. Regarding the random slope, we see that there is a significant average of γ_{20}^{HAR}, and the intercept of β_{2j}^{HAR} is estimated to

---

[6] Doing so would violate the limitations on the number of variables allowed in the demo version of *Mplus*, which we use for practical examples throughout the book. Therefore, in this model, we do not estimate the contextual effects of salary on perceived skills.

$$\begin{cases} HW1_{ij} = \mu_{00}^{HW1} + \mu_{02}^{HW1\ HAB}\gamma_{00} + \mu_{02}^{HW1\ HAB}\gamma_{01}NEM_j + \mu_{02}^{HW1\ HAB\ REB}\gamma_{02}\gamma_{00} + \mu_{02}^{HW1\ HAB\ REB}\gamma_{02}\gamma_{01}NEM_j + \mu_{02}^{HW1\ HAB\ REB}\gamma_{02}\upsilon_{0j} + \mu_{02}^{HW1\ HAB}\upsilon_{0j} + \\ \upsilon_{0j}^{HW1} + \mu_{02}^{HW1\ HAR}\beta_{0j} + \mu_{02}^{HW1\ HAR}\beta_1 RES_{ij} + \mu_{02}^{HW1\ HAR}\gamma_{20}PAY_{ij} + \mu_{02}^{HW1\ HAR}\gamma_{21}PAY_{ij}NEM_j + \mu_{02}^{HW1\ HAR}\upsilon_{2j}PAY_{ij} + \mu_{02}^{HW1}\varepsilon_{ij} + \varepsilon_{ij}^{HW1} \\ HW2_{ij} = \mu_{00}^{HW2} + \mu_{02}^{HW2\ HAB}\gamma_{00} + \mu_{02}^{HW2\ HAB}\gamma_{01}NEM_j + \mu_{02}^{HW2\ HAB\ REB}\gamma_{02}\gamma_{00} + \mu_{02}^{HW2\ HAB\ REB}\gamma_{02}\gamma_{01}NEM_j + \mu_{02}^{HW2\ HAB\ REB}\gamma_{02}\upsilon_{0j} + \mu_{02}^{HW2\ HAB}\upsilon_{0j} + \\ \upsilon_{0j}^{HW2} + \mu_{02}^{HW1\ HAR}\beta_{0j} + \mu_{02}^{HW2\ HAR}\beta_1 RES_{ij} + \mu_{02}^{HW2\ HAR}\gamma_{20}PAY_{ij} + \mu_{02}^{HW2\ HAR}\gamma_{21}PAY_{ij}NEM_j + \mu_{02}^{HW2\ HAR}\upsilon_{2j}PAY_{ij} + \mu_{02}^{HW2\ HAR}\varepsilon_{ij} + \varepsilon_{ij}^{HW2} \\ HW3_{ij} = \mu_{00}^{HW3} + \mu_{02}^{HW3\ HAB}\gamma_{00} + \mu_{02}^{HW3\ HAB}\gamma_{01}NEM_j + \mu_{02}^{HW3\ HAB\ REB}\gamma_{02}\gamma_{00} + \mu_{02}^{HW3\ HAB\ REB}\gamma_{02}\gamma_{01}NEM_j + \mu_{02}^{HW3\ HAB\ REB}\gamma_{02}\upsilon_{0j} + \mu_{02}^{HW3\ HAB}\upsilon_{0j} + \\ \mu_{02}^{HW1\ HAR}\beta_{0j} + \mu_{02}^{HW3\ HAR}\beta_1 RES_{ij} + \mu_{02}^{HW3\ HAR}\gamma_{20}PAY_{ij} + \mu_{02}^{HW3\ HAR}\gamma_{21}PAY_{ij}NEM_j + \mu_{02}^{HW3\ HAR}\upsilon_{2j}PAY_{ij} + \mu_{02}^{HW3\ HAR}\varepsilon_{ij} + \varepsilon_{ij}^{HW3} \end{cases} \quad (4.18)$$

Figure 4.8 Random Slopes Model (unstandardized results)

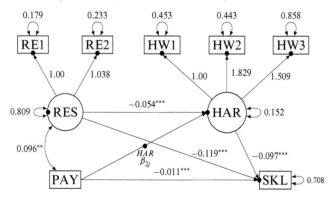
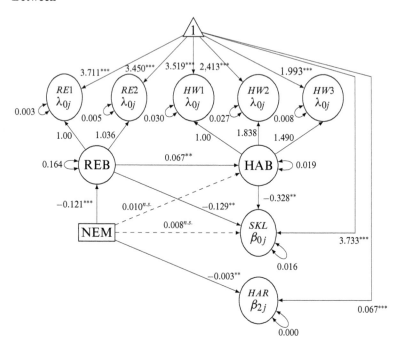

be 0.067. That is virtually the same as the estimated effect of salary on job difficulty at the within level in the previous model, from Figure 4.6, which was 0.065. However, we observe a small Level 2 variance of the slopes (due to the small scale, it is rounded to zero), which is significantly predicted by the number of employees in a company (-0.003^*). This means that the more workers there are in a company, the weaker is the relationship between higher salaries and higher job difficulty. In a company with thousands of employees, high pay has a smaller relationship with perception of a hard job than in a company with a dozen workers. Other estimates have remained the same as they were in the previous model.

Model Comparison

Finally, we can use comparative fit statistics to contrast the performance of all models tested in this chapter. The AIC and the BIC for each model are in Table 4.2. For the first four models, where absolute fit indices are available, we see good fit. Models with random slopes, however, have no χ^2-based fit indicators, and so we base the assessment only on comparative indices.

In Table 4.2, we see a continuous improvement in model fit from Models 1 through 4. Both the AIC and the BIC get smaller. The first model allows only the perceived skill (SKL) to have both its within- and between-level variances estimated, while the second does so for both the skill (SKL) and the salary (PAY) variables. For all five indicators, the Level 2 variance component was fixed to 0. However, since all variables measured at the employee level also display at least a bit of cross-company variation,[7] it was expected that Models 3 and 4 would have

Table 4.2 Comparative Model Fit

Order	Model	AIC	BIC	Nested in
1	Random Intercept (Fig. 4.2)	391,916	392,173	2, 3, 4, 5
2	Latent Covariate (Fig. 4.3)	391,290	391,541	3, 4
3	Doubly Latent I (Fig. 4.5)	388,147	388,524	—
4	Doubly Latent II (Fig. 4.6)	388,141	388,494	3
5	Random Slopes (Fig. 4.8)	388,806	389,089	—

---

[7] The ICCs for the five indicators range from 0.061 ($HW3$) to 0.173 ($RE1$).

better fit, considering that all indicators have both variance components estimated. In this sense, it is interesting to note that the much more parsimonious fourth model, in which we fix factor loadings to be the same in both levels, has a superior fit than the third model, which is exactly the same but with factor loadings free to vary. Despite an equality constraint, model fit still shows a slight improvement based on the AIC and BIC.

Because we dropped β_{0j}^{PAY} from the random slopes model, the only model that is nested in it is the very first. Understandably, the random slopes model has better fit, since it allows for many more parameters to be freely estimated. A direct comparison to the more similar Models 3 and 4 show worse fit for the random slopes option, suggesting it might not be necessary to allow that slope to vary across groups. However, it is arguable whether the AIC and the BIC can be used to compare nonnested models, even in cases when sample size and the specific cases match.

Last, a word of caution that the model comparisons above are possible because we have used listwise deletion to deal with missing data, whereby any observation that had a missing entry for any variable used in any of the models has been removed. The sample size is stable at 18,918 employees and 1,723 companies. With other treatments of missing data, or if different variables are used in each model, model fit comparisons are something the researcher must pay attention to.

Regarding sample sizes, it is also important to note that, besides the Level 2 sample size discussions typical of multilevel modeling, in MSEM, it is recommended that the number of clusters be larger than the number of parameters estimated in a model. While typical software will estimate models with more free parameters than Level 2 units, these are subject to empirical underidentification, so that estimated parameters, as well as standard errors, might not be reliable. This is less of a problem with data structures such as that seen in this chapter, with more than 1,000 companies. It becomes a potential issue in studies with cross-national surveys, for example. If we have an EU-wide one, the Level 2 "sample" size of 25 to 28 does not allow for even moderately large models without running into having more estimated parameters than clusters.

Summary

This chapter combines the multilevel path models of Chapter 2 with the multilevel latent variable models of Chapter 3 rounding out the bulk of the knowledge we were hoping to pass along on MSEM. The examples work through some of the possibilities offered by multilevel structural equation models with multi-indicator latent variables. We started with a simple design, with a single observed outcome at both levels, and advanced with latent contextual models, decomposing the variance of an exogenous variable and adding Level 2 outcome and latent variables. Last, we demonstrate a random slopes model with between-level latent variables and discussed a few limitations imposed by the estimation techniques.

These models can be expanded in several ways. For example, if one identifies breakdowns in measurement invariance of an instrument, it is possible to have a contextual covariate that explains why a certain item is noninvariant in some groups (Davidov, Dülmer, Schlüter, Schmidt, & Meuleman, 2012). Furthermore, all the models discussed here can be extended into applications with categorical outcomes, as well as alternative estimation methods such as a Bayesian framework. We present some alternatives in greater detail in the conclusion.

CHAPTER 5. CONCLUSION

Structural equation modeling and multilevel modeling have developed side by side to tackle quite different modeling problems. SEM's primary purpose is to reproduce the data covariance matrix. MLM deals with regression situations where the data structure is clustered and hence violates the independence of observations assumptions. Once one gets into advanced topics, however, SEM's flexibility as a modeling approach is usually limited more by the researcher's imagination, who boxes the technique into the original intended purpose driven by path dependence. This poses a potential pitfall for people who would like to venture deeper into the MSEM scholarship. We therefore begin with some guidance that will help the reader navigate and follow up, by pointing to the more advanced (and, sometimes, future) topics on MSEM we did not cover.

Initial attempts to marry the SEM and MLM literature did not approach the relationship like L. K. Muthén and Muthén (1998–2017), and now this volume, by using what MLM gave to regression (the random effect) and applying it to SEM path and intercept (mean structure) estimates. Instead, these attempts reformulated SEM in ways that deal with multilevel data structures. In fact, SEM has the flexibility to specify multilevel models (including all the models presented in this volume) as a single-level structural equation model. This reality understandably causes much confusion in the MSEM world. Anyone who wants to venture into the technical literature on MSEM will not be able to avoid the two completely different lines of thinking.

The first encounter with the two approaches will probably emerge as one starts to think about two different formulations of a growth model specified to test trajectories of change within individuals.[1] A multilevel modeler would surely specify such a model as a within-person observation design where person is the Level 2 and the multiple observations within the person are at Level 1. The main independent variable of interest is the passage of time (for more information, see Singer & Willet, 2003, chap. 1–7). A structural equation modeler would set this model up as a latent growth curve (Preacher, Wichman, MacCallum, & Briggs,

---

[1] Growth models receive their name from modeling the actual change in height—growth—of children. The approach is common in education research for modeling learning over time.

2008), which is a single-level structural equation model that is actually equivalent to the multilevel model described here. These two different specifications in modeling multilevel data structures already cause much confusion in the literature.[2] The situation is even more complicated. Moving beyond simple growth curve models but applying the same logic, every SEM can be turned into a multilevel model, not just the ones modeling change. How these models work is beyond the scope of this book. For a simple introduction, see Mehta and Neale (2005). This is the logic implemented in Mehta's xxM for R, and this is the logic that was predominant in the early days of MSEM. Today, this approach is sometimes referred to as NL-SEM, or nSEM, short for N-level structural equation modeling, as it is a technique that can turn any single-level structural equation model into a multilevel structural equation model with any number of levels. Despite its high threshold for entry, in terms of statistical background, we strongly recommend to readers who wish to get a full picture of the MSEM landscape to consult this other strand as well. If we have avoided mentioning it so far, it is only because our goal with the book was to make this difficult subject more accessible. Anyone trying to move beyond this introduction may come across these more complex formulations. If they remain unfazed, we recommend that the initiation process to this other logic start with Mehta and Neale (2005).

Moving beyond the two worlds of MSEM and back into the world presented in this book, it is important to take stock of the strengths, limitations, and trade-offs modelers have to consider when applying MSEM. On one hand, MSEM allows researchers to pose novel questions and to draw on the strengths of both MLM and SEM approaches when attempting to answer them. These strengths include the ability to measure latent constructs and to do so across group contexts such as schools, firms, or countries, as well as to overcome limitations imposed by imbalanced sample sizes across groups. On the other hand, adopting this framework increases analytic and computational complexity and places a high theoretical burden on the researcher, who now must produce and justify plausible theoretical causal mechanisms that are to be tested with MSEM. Social scientists often do not have a good track record in this respect, even with regard to the comparatively simple multilevel

---

[2] Before fully implementing three-level models, L. K. Muthén and Muthén (1998–2017) claimed in the user's guide that *Mplus* can, indeed, run three-level models by fitting a latent growth curve model and layering a second level on top.

regression models. In this concluding chapter, we also suggest a few additional directions in which an eager consumer of the MSEM framework can venture into, after absorbing the material presented in the book.

Throughout the book, our statistical specifications have tried to explain continuous outcomes. In the terminology commonly used in generalized linear modeling (GLM), our models have made use of the *identity* link function. Here, we are directly modeling the mean of the outcome, conditional on the k covariates in the specification.

$$\mu = \sum_{i=1}^{k} x_i \beta_i \qquad (5.1)$$

Even though a number of phenomena can be studied through this specification, this pales in comparison to the ones that are recorded on categorical scales, be they ordered or unordered. One's sense of political efficacy or satisfaction with democracy, preferences for redistribution, or attitudes toward gender relations are all orientations typically measured using scales with 5, 7, or, at most, 11 categories. Even more so, behaviors such as turnout in elections, the decision to emigrate, or casting a vote for a populist party are recorded on dichotomous scales, with a yes/no choice. Rhemtulla, Brosseau-Liard, and Savalei (2012) tested this issue in SEM and concluded that variables with five categories can be treated as continuous (unless the data-generating mechanism clearly and visibly deviates from normality). We made sure to follow this principle in the selection of our examples as we believe (but have not tested) that this result generalizes to MSEM. People uncomfortable with this assumption or with ordered variables with fewer than five categories, however, also have options.

For these instances, the identity link function is no longer suitable. More complex functions, which act as "translators" between the scale on which the outcome is measured and the *linear predictor* produced by the linear additive modeling specification, are needed. Such common functions are the *logit*, $log\{\mu/(1-\mu)\}$; the *probit*, $\Phi^{-1}(\mu)$; or the *complementary log-log* function, $log\{-log(1-\mu)\}$ (McCullagh & Nelder, 1989, chap. 2).[3] Their use in MLM is by now firmly established (Hox, 2010, chaps. 6–7), and the reader should feel heartened by the knowledge

---

[3] In the case of the probit, Φ designates the cumulative distribution function (CDF) for the Gaussian distribution.

that the MSEM framework covered so far can be extended to accommodate them as well (Rabe-Hesketh, Skrondal, & Pickles, 2004). At the same time, we advise caution when advancing to the *generalized linear latent and mixed modeling* (GLLAMM) framework.[4] There is limited (simulation-based) work exploring the conditions under which such models can produce sound estimates, particularly when we contrast this with the multilevel framework (e.g., McNeish & Stapleton, 2016) or more basic MSEM specifications (Hox & Maas, 2001). As a result of this, the reader would venture out with minimal guidance into as yet unexplored territory.

Up to this point, we have, quite explicitly, avoided discussions of weighting. For the educational purposes of the examples presented in this volume, simplicity and broad and software-independent reproducibility was our primary guiding principle. But, in fact, it is quite inappropriate to ignore potential sampling weights in most, if not all, of the examples presented. So let's consider how weights, usually as provided by the organizations that generated the data, could be incorporated. First, Stapleton (2002) applauded the use of MSEM as a potential way to overcome the biases associated with a clustered sampling design.[5] The method gives us the opportunity to incorporate the clustering in the sampling design directly. While this approach can overcome the biases associated with the clustering, under- and overrepresentation of certain individuals or even higher-level units could produce strong biases if the relevant variables are not accounted for in the model or weights are not applied. Fortunately, some application of weights is already available (Asparouhov, 2006) and implemented (L. K. Muthén & Muthén, 1998–2017). Asparouhov (2006), despite also being highly technical, offers in the conclusion an easy-to-follow, step-by-step guide on how to proceed with weighting for MSEM.

The modeling framework presented here could also be extended into situations where the data are clustered at more levels of analysis (Preacher, 2011; Steele, Clarke, Leckie, Allan, & Johnston, 2017). Typical situations might be school settings, with students clustered in classrooms and then further in schools (Palardy, 2008), or electoral

---

[4] Chapter 3 also briefly discussed an application of ordinal data to multilevel confirmatory factor models. For more information, see Grilli and Rampichini (2007).

[5] The article, although quite technical, extended well beyond this proposition and suggested a number of ways different types of weights can be incorporated into the estimation of the variance covariance matrices used in MSEM.

settings, with voters clustered in electoral districts, further clustered in counties, states, or cantons. While it is easy to see additional levels of analysis in various data sources, readers are warned before venturing into the specification of these levels. As a rule of thumb, if there are no variables available for a level of analysis or no theory linking covariates at this level with outcomes, the autocorrelation stemming from the clustering can be accounted for by simpler methods, such as clustered standard errors or fixed effects.[6] As another general rule, if the sample size on the highest level is below what we would be comfortable with for ordinary least squares (OLS) regression, or we doubt the potential for a normal distribution emerging in the residuals because of the presence of such few cases, it is simply not a good idea to specify a level of analysis.

We also remind the reader that adding yet another level of analysis represents an additional complication not only in terms of notation or estimation. Three-level structures bring with them the possibility of specifying associations between variables measured at any of the three levels, which fragments the variance of any variable across the levels and decreases the power to detect a significant relationship. Severe imbalances between group sizes at the higher levels might bias the estimates in unpredictable ways, and much empirical work remains to be done in understanding this phenomenon as well.

A researcher coming from the MLM tradition might also be familiar with data structures that are decidedly nonhierarchical. In such instances, observations are simultaneously allocated to multiple grouping structures, with no possibility of assigning a clear hierarchical order to these structures. A few typical situations are suggested by Goldstein (2011): Students can be clustered both in schools and in neighborhoods, producing a cross-classified structure of Neighborhood × School; longitudinal measures are used for individual-level attitudes, with answers clustered both in time points and interviewer (assuming that multiple interviewers are used over the course of the survey waves). Exogenous covariates can be used at both school and neighborhood levels, or time and interviewer levels, while the residual variance can now be decomposed into, often overlapping, variances at each level of the grouping structure (e.g., one for the school level and one for the neighborhood

---

[6] Although clustered standard errors, to date, are only sparsely implemented in software and the inclusion of fixed effects in structural equation models is not straightforward.

level). While the use of such data structures is established in the MLM framework, for MSEM models, they still represent an emerging topic (see González, De Boeck, & Tuerlinckx, 2008; Nestler & Back, 2017). There might be cases in which there is grouping at the lower level: For example, students are clustered within schools but are also split into boys and girls. A pure multigroup MSEM approach would consider boys and girls from the same school as independent observations from one another, violating the hierarchical structure of the data (Ryu & Mehta, 2017).

Researchers with these kinds of data have currently two alternatives. The first, with a maximum likelihood approach, is rooted in the aforementioned n-level SEM (Mehta, 2013a). It involves restructuring the data into *levels*, each one being one clustering option (say, boys, girls, and schools), and defining the relationships between them before setting up the model to be fit. Readers are referred to Mehta (2013a, 2013b) for technical details and implementation. The second alternative uses Bayesian estimation (Asparouhov & Muthén, 2015), with its accompanying computational demands in the case of large data sets and many random effects to be estimated. We advise those who want to analyze such cross-classified data structures to only do so after a crash course in Bayesian methods, at the level presented in Gill (2015) or Gelman et al. (2014).

And even beyond three (or more) levels and cross-classified models, other forms of grouping have already received attention in the multilevel modeling literature. Multiple membership models allow each Level 1 unit to be a member in multiple Level 2 units that are not mutually exclusive (Fielding & Goldstein, 2006; Goldstein, 2011). For example, students could have different educational profiles as they progress through the secondary cycle: Some students attend the same institution, while others move between two or even three institutions. In an internship program, candidates might be rotated between multiple teams in a company: Some candidates might be embedded for the entire duration of the internship in one team, while others might be exposed to the climate of multiple teams. The challenge for these models is coming up with a set of weights that quantify the influence of each school or work team for each person. A natural start would be to use as weights the share of time spent in the school or team (Fielding & Goldstein, 2006, p. 34). An intern who spends all his or her time in a team might get a weight of 1 for this team and 0 for all other teams. Another intern who spent equal time in two teams might get 0.5 for each of these two teams and 0 for the rest. These very same research examples could also accommodate questions that refer to 2 → 1 → 1 mediation configurations (e.g., the impact of school average ability on individual-level career aspirations) and how this is moderated by individual-level self-concept (Nagengast & Marsh, 2012). However, such models have not yet been explored in the MSEM literature, although they certainly hold much promise for the future of this field.

Those who come from educational studies are familiar with multitrait multimethod (MTMM) analysis (Campbell & Fiske, 1959). This approach is often used, for example, when multiple students are evaluating teachers on a variety of traits. It allows for the separation of method (e.g., question style), trait, and error components, giving a rigorous assessment of measurement techniques. Often, however, MTMM uses a single combination of method per trait—for instance, only one question, with one style, measures each trait (Kenny, 1976; Marsh, 1993). As the reader might imagine at this point, such a structure might be suitable for multilevel modeling. More specifically, Eid et al. (2008) show that if raters are interchangeable (e.g., randomly drawn students), MTMM with multiple indicators per trait method can be specified as a multilevel CFA, in which case raters are the Level 1, in which we model the method factor(s), and Level 2 are the targets (say, teachers), where we model the trait latent variables.

Another important topic within SEM is multigroup analysis. It is used to compare groups, test measurement invariance, and assess treatment effects from experimental and quasi-experimental studies, among other applications. It is possible to have a multigroup-multilevel SEM, in which clusters j are sampled from different categories of a population g (B. O. Muthén, Khoo, & Gustafsson, 1997). Imagine, for instance, a typical data-nesting structure of students within schools. However, suppose that there are two kinds of schools, public and private; this is the case with the PISA data used in Chapter 3. We are interested in finding out whether the relationship between interest and performance (both modeled as latent variables with multiple indicators) is the same across the two types of schools. This comparison can be done by fitting a multiple-group MSEM in which one or more paths are allowed to vary (or not) across the two groups. The difference between the model with and without the equality restriction set for the paths of interests is evaluated with a simple model comparison test. For the mathematical description of such models, please consult B. O. Muthén et al. (1997); B. O. Muthén (2002), while Mayer et al. (2014) provide a didactic example.

Missing data in multilevel models has received some attention over the years.[7] The first class of solutions is imputation based. Multilevel data can be represented as single-level data with the inclusion of dummy variables for each clustering unit (Graham, 2009),[8] although this approach can provide distorted parameter estimates and standard errors (Andridge, 2011; Drechsler, 2015; Enders, Mistler, & Keller, 2016; Lüdtke, Robitzsch, & Grund, 2017). Two approaches overcome this problem by specifying a multilevel model for the imputation. The older one, known as joint modeling, uses a Bayesian framework (Schafer, 2001; Schafer & Yucel, 2002) to estimate a multilevel imputation model and sample from the posterior of the missing values in a single step. It is mathematically elegant and unbiased but limited in its ability to address real-life research problems like categorical data, imputing for random slopes models, missing values in exogenous variables, or on higher levels of the analysis (Grund, Lüdtke, & Robitzsch, 2016). The other approach, called fully conditional specification or chained equations, redefines all variables with missing as outcomes and iteratively imputes based on the predicted values of these (multilevel) regression models. While certainly less elegant, it has flexibility to consider dichotomous or ordinal variables, partition relationships across Level 1 and Level 2 effects already in the imputation stage, and account for incomplete Level 2 variables. For a flexible implementation of this approach, see Enders, Keller, and Levy (2017).

Neither imputation procedure can consider random slopes flexibly when data are missing on exogenous variables. Graham (2012) suggests that imputations may need to be done within the clusters in these situations, while Schafer (2001) proposes to accept the bias for slope variance that stems from making these variables endogenous.[9] It is important to note that although theoretically these imputation approaches extend naturally to MSEM, no research to date has tested their unbiasedness and efficiency in the MSEM context. Also, while the combination of multiply imputed parameter estimates is the same independent of modeling approach, combining other statistics (like model fit) is potentially

---

[7] For an overview of the missing data problem for hierarchical data, see van Buuren (2011).
[8] For an implementation of this approach, see the time-series cross-sectional models in Honaker and King (2010).
[9] See also Grund et al. (2016).

less straightforward and certainly understudied even in the single-level SEM context.[10]

Structural equation modelers are more used to applying full-information maximum likelihood estimation procedures to missing data and, given that every multilevel model can be rewritten as a structural equation model, these approaches should extend naturally as well. While implementations exist (L. K. Muthén & Muthén, 1998–2017), documentation of these implementations is scarce, and tests or comparisons of their applications are, to date, nonexistent as far as the authors can tell. Also, while missing data approaches for Level 1 variables are plentiful, with Level 2 being more scarce, approaches to deal with missing data on group membership are also practically nonexistent.[11] All of these holes in the literature are fertile ground for methodological research.

Finally, one important characteristic of structural equation models (vs. other techniques popular in the social sciences) is a careful consideration of model fit. While almost all published SEM analyses present some evidence against the fit and consequent interpretability of the model (usually in form of a significant χ^2 test, usually dismissed as nothing more than accumulated minor deviations in structure and normality stemming from a larger sample size), SEM is, at least, careful to consider a slew of statistics that, one way or another, are designed to assess the fit of the model. Multilevel modeling, from its days of inception, took the question of fit very lightly. Model fit is only considered in comparison to more parsimonious models (sometimes ones lacking covariates entirely). Needless to say, any variance explained (especially in large samples) will show improvement of fit in these instances, but mechanisms are almost entirely unavailable for assessing overall fit of multilevel models. The merging of the two modeling approaches certainly poses the risk of swaying SEM, a modeling approach careful with the presentation of reliable overall models from the perspective of model fit, toward the approach of multilevel models, a paradigm that does not care. It would be a shame to see SEM go the MLM route as the modeling of multilevel data structures becomes a part of the SEM paradigm. Unfortunately, signs of this

---

[10] See Littvay (2009) for an example of how missing data can cause issues even with full-information maximum likelihood estimation.

[11] However, van Buuren (2011) suggests imputation could be used potentially. The author believes mixture modeling approaches would be more fruitful, but the details around implementation certainly need development.

change are already present. Early software that allowed for the structural equation modeling of multilevel data structures did not offer the kind (and variety) of fit statistics that structural equation modelers are used to considering. The reasons for this are less normative and more due to technical limitations; nonetheless, it is swaying the field in the direction of multilevel modeling where fit is not considered. Normatively speaking, we need to ensure that the MSEM paradigm keeps the best, and not the worst, aspects of the two modeling approaches it is merging. And this means a careful consideration of model fit.

REFERENCES

Akaike, H. (1973). Information theory and an extension of the maximum likelihood principle. In B. N. Petrov & F. Csaki (Eds.), *Second International Symposium on Information Theory* (pp. 267–281). Budapest: Akademiai Kiado.

Andridge, R. R. (2011). Quantifying the impact of fixed effects modeling of clusters in multiple imputation for cluster randomized trials. *Biometrical Journal, 53*(1), 57–74.

Asparouhov, T. (2006). General multi-level modeling with sampling weights. *Communications in Statistics: Theory and Methods, 35*(3), 439–460.

Asparouhov, T., & Muthén, B. O. (2015). General random effect latent variable modeling: Random subjects, items, contexts, and parameters. In J. R. Harring, L. M. Stapleton, & S. N. Beretvas (Eds.), *Advances in multilevel modeling for educational research: Addressing practical issues found in real-world applications*. Charlotte, NC: Information Age Publishing, Inc.

Baron, R. M., & Kenny, D. A. (1986). The moderator–mediator variable distinction in social psychological research: Conceptual, strategic, and statistical considerations. *Journal of Personality and Social Psychology, 51*(6), 1173–1182.

Barrett, P. (2007). Structural equation modelling: Adjudging model fit. *Personality and Individual Differences, 42*(5), 815–824.

Bauer, D. J., Preacher, K. J., & Gil, K. M. (2006). Conceptualizing and testing random indirect effects and moderated mediation in multilevel models: New procedures and recommendations. *Psychological Methods, 11*(2), 142–163.

Blau, P. M., & Duncan, O. D. (1967). *The American occupational structure*. New York: John Wiley & Sons.

Bollen, K. A. (1987). Total, direct, and indirect effects in structural equation models. *Sociological Methodology, 17*, 37–69.

Bollen, K. A. (1989). *Structural equations with latent variables*. New York: Wiley–Interscience.

Bollen, K. A., & Stine, R. A. (1992). Bootstrapping goodness-of-fit measures in structural equation models. *Sociological Methods Research, 21*(2), 205–229.

Boyd, L. H., & Iversen, G. R. (1979). *Contextual analysis: Concepts and statistical techniques*. Belmont, CA: Wadsworth.

Burnham, K. P., & Anderson, D. R. (2002). *Model selection and multimodel inference: A practical information-theoretic approach* (2nd ed.). New York: Springer.

Campbell, D. T., & Fiske, D. W. (1959). Convergent and discriminant validation by the multitrait multimethod matrix. *Psychological Bulletin, 56*(2), 81–105.

Davidov, E., Dülmer, H., Schlüter, E., Schmidt, P., & Meuleman, B. (2012). Using a multilevel structural equation modeling approach to explain cross-cultural measurement noninvariance. *Journal of Cross-Cultural Psychology, 43*(4), 558–575.

De Boeck, P. (2008). Random item IRT models. *Psychometrika, 73*(4), 533–559.

De Jong, M. G., Steenkamp, J.-B. E. M., & Fox, J.-P. (2007). Relaxing measurement invariance in cross-national consumer research using a hierarchical IRT model. *Journal of Consumer Research, 34*(2), 260–278.

Drechsler, J. (2015). Multiple imputation of multilevel missing data: Rigor versus simplicity. *Journal of Educational and Behavioral Statistics, 40*(1), 69–95.

Duncan, O. D. (1968). Ability and achievement. *Biodemography and Social Biology, 15*(1), 1–11.

Duncan, O. D., Haller, A. O., & Portes, A. (1971). Peer influences on aspirations: A reinterpretation. In H. M. Blalock, Jr. (Ed.), *Causal models in the social sciences* (pp. 219–244). London: Macmillan.
Eid, M., Nussbeck, F. W., Geiser, C., Cole, D. A., Gollwitzer, M., & Lischetzke, T. (2008). Structural equation modeling of multitrait-multimethod data: Different models for different types of methods. *Psychological Methods, 13*(3), 230–253.
Elff, M., Heisig, J., Schaeffer, M., & Shikano, S. (2016). *No need to turn Bayesian in multilevel analysis with few clusters: How frequentist methods provide unbiased estimates and accurate inference* (Working Paper). College Park, MA: SocArXiv. doi: 10.17605/OSF.IO/Z65S4
Eliason, S. R. (1993). *Maximum likelihood estimation: Logic and practice*. Thousand Oaks, CA: SAGE.
Enders, C. K. (2010). *Applied missing data analysis*. New York: Guilford.
Enders, C. K., Keller, B. T., & Levy, R. (2017). A fully conditional specification approach to multilevel imputation of categorical and continuous variables. *Psychological Methods, 23*(2), 298–317.
Enders, C. K., Mistler, S. A., & Keller, B. T. (2016). Multilevel multiple imputation: A review and evaluation of joint modeling and chained equations imputation. *Psychological Methods, 21*(2), 222–240.
Enders, C. K., & Tofighi, D. (2007). Centering predictor variables in cross-sectional multilevel models: A new look at an old issue. *Psychological Methods, 12*(2), 121–138.
Epstein, D. L., Bates, R., Goldstone, J., Kristensen, I., & O'Halloran, S. (2006). Democratic transitions. *American Journal of Political Science, 50*(3), 551–569.
Fielding, A., & Goldstein, H. (2006). *Cross-classified and multiple membership structures in multilevel models: An introduction and review* (Technical Report). London: Institute of Education, University College. Retrieved from http://dera.ioe.ac.uk/6469/1/RR791.pdf
Gelman, A., Carlin, J. B., Stern, H. S., Dunson, D. B., Vehtari, A., & Rubin, D. B. (2014). *Bayesian data analysis* (3rd ed.). Boca Raton, FL: Chapman & Hall/CRC.
Gelman, A., & Hill, J. (2007). *Data analysis using regression and multilevel/hierarchical models*. New York: Cambridge University Press.
Gill, J. (2015). *Bayesian methods: A social and behavioral sciences approach* (3rd ed.). Boca Raton, FL: Chapman & Hall/CRC.
Goldstein, H. (2011). *Multilevel statistical models* (4th ed.). New York: John Wiley.
Goldstein, H., & McDonald, R. P. (1988). A general model for the analysis of multilevel data. *Psychometrika, 53*(4), 455–467.
González, J., De Boeck, P., & Tuerlinckx, F. (2008). A double-structure structural equation model for three-mode data. *Psychological Methods, 13*(4), 337–353.
Goodin, R., & Dryzek, J. (1980). Rational participation: The politics of relative power. *British Journal of Political Science, 10*(3), 273–292.
Goodman, L. A. (1960). On the exact variance of products. *Journal of the American Statistical Association, 55*(292), 708–713.
Graham, J. W. (2009). Missing data analysis: Making it work in the real world. *Annual Review of Psychology, 60*(1), 549–576.
Graham, J. W. (2012). *Missing data: Analysis and design*. New York: Springer.
Grilli, L., & Rampichini, C. (2007). Multilevel factor models for ordinal variables. *Structural Equation Modeling: A Multidisciplinary Journal, 14*(1), 1–25.
Grund, S., Lüdtke, O., & Robitzsch, A. (2016). Multiple imputation of multilevel missing data: An introduction to the R package pan. *SAGE Open, 6*(4), 1–17.

Hancock, G. R., & Mueller, R. O. (2006). *Structural equation modeling: A second course*. Greenwich, CT: Information Age Publishing.

Hayduk, L. A. (1987). *Structural equation modeling with LISREL: Essentials and advances*. Baltimore, MD: Johns Hopkins University Press.

Hayduk, L. A., Cummings, G., Boadu, K., Pazderka-Robinson, H., & Boulianne, S. (2007). Testing! testing! one, two, three: Testing the theory in structural equation models! *Personality and Individual Differences*, *42*(5), 841–850.

Hayduk, L. A., & Littvay, L. (2012). Should researchers use single indicators, best indicators, or multiple indicators in structural equation models? *BMC Medical Research Methodology*, *12*, 159.

Hayes, A. F. (2013). *Introduction to mediation, moderation, and conditional process analysis: A regression-based approach*. New York: Guilford.

Heck, R. H., & Thomas, S. L. (2015). *An introduction to multilevel modeling techniques: MLM and SEM approaches using Mplus* (3rd ed.). New York: Routledge.

Honaker, J., & King, G. (2010). What to do about missing values in time-series cross-section data. *American Journal of Political Science*, *54*(2), 561–581.

Hox, J. J. (2010). *Multilevel analysis: Techniques and applications* (2nd ed.). New York: Routledge.

Hox, J. J., & Maas, C. J. M. (2001). The accuracy of multilevel structural equation modeling with pseudobalanced groups and small samples. *Structural Equation Modeling: A Multidisciplinary Journal*, *8*(2), 157–174.

Hox, J. J., & Roberts, J. K. (Eds.). (2011). *Handbook of advanced multilevel analysis*. New York: Routledge.

Hoyle, R. H. (2012). *Handbook of structural equation modeling*. New York: Guilford.

Iacobucci, D. (2008). *Mediation analysis*. Thousand Oaks, CA: SAGE.

Inglehart, R. F., & Baker, W. E. (2000). Modernization, cultural change, and the persistence of traditional values. *American Sociological Review*, *65*(1), 19–51.

Inglehart, R. F., & Welzel, C. (2009). How development leads to democracy: What we know about modernization. *Foreign Policy*, *88*(2), 33–48.

Jöreskog, K. G. (1973). A general method for estimating a linear structural equation system. In A. S. Goldberger & O. D. Duncan (Eds.), *Structural equation models in the social sciences* (pp. 85–112). New York: Seminar.

Kaplan, D., & Depaoli, S. (2012). Bayesian structural equation modeling. In R. H. Hoyle (Ed.), *Handbook of structural equation modeling* (pp. 650–673). New York: Guilford.

Keesling, J. W. (1972). *Maximum likelihood approaches to causal flow analysis* (Doctoral dissertation). University of Chicago.

Kenny, D. A. (1976). An empirical application of confirmatory factor analysis to the multitrait-multimethod matrix. *Journal of Experimental Social Psychology*, *12*(3), 247–252.

Kenny, D. A., Korchmaros, J. D., & Bolger, N. (2003). Lower level mediation in multilevel models. *Psychological Methods*, *8*(2), 115–128.

Kline, R. B. (2015). *Principles and practice of structural equation modeling* (4th ed.). New York: Guilford.

Kozlowski, S. W. J., & Klein, K. J. (2000). A multilevel approach to theory and research in organizations: Contextual, temporal, and emergent processes. In K. J. Klein & S. W. J. Kozlowski (Eds.), *Multilevel theory, research, and methods in organizations: Foundations, extensions, and new directions* (pp. 3–90). San Francisco, CA: Jossey-Bass.

Kraemer, H. C., Kiernan, M., Essex, M., & Kupfer, D. J. (2008). How and why criteria defining moderators and mediators differ between the Baron & Kenny and MacArthur approaches. *Health Psychology, 27*(2), S101–S108.
Kraemer, H. C., Wilson, G. T., Fairburn, C. G., & Agras, W. S. (2002). Mediators and moderators of treatment effects in randomized clinical trials. *Archives of General Psychiatry, 59*(10), 877–883.
Kreft, I. G. G. (1996). *Are multilevel techniques necessary? An overview, including simulation studies*. Unpublished manuscript. Los Angeles: California State University.
Kreft, I. G. G., & de Leeuw, J. (1998). *Introducing multilevel modeling*. London: SAGE.
Kreft, I. G. G., de Leeuw, J., & Aiken, L. S. (1995). The effect of different forms of centering in hierarchical linear models. *Multivariate Behavioral Research, 30*(1), 1–21.
Krull, J. L., & MacKinnon, D. P. (2001). Multilevel modeling of individual and group level mediated effects. *Multivariate Behavioral Research, 36*(2), 249–277.
Kruschke, J. K. (2014). *Doing Bayesian data analysis: A tutorial with R, JAGS, and Stan*. London: Academic Press.
Little, T. D. (2013). *Longitudinal structural equation modeling*. New York: Guilford.
Littvay, L. (2009). Questionnaire design considerations with planned missing data. *Review of Psychology, 16*(2), 103–114.
Lüdtke, O., Marsh, H. W., Robitzsch, A., Trautwein, U., Asparouhov, T., & Muthén, B. (2008). The multilevel latent covariate model: A new, more reliable approach to group-level effects in contextual studies. *Psychological Methods, 13*(3), 203–229.
Lüdtke, O., Robitzsch, A., & Grund, S. (2017). Multiple imputation of missing data in multilevel designs: A comparison of different strategies. *Psychological Methods, 22*(1), 141–165.
Luke, D. A. (2004). *Multilevel modeling*. Thousand Oaks, CA: SAGE.
Maas, C. J. M., & Hox, J. J. (2005). Sufficient sample sizes for multilevel modeling. *Methodology, 1*(3), 86–92.
Marsh, H. W. (1993). Multitrait-multimethod analyses: Inferring each trait-method combination with multiple indicators. *Applied Measurement in Education, 6*(1), 49–81.
Marsh, H. W., Lüdtke, O., Robitzsch, A., Trautwein, U., Asparouhov, T., Muthén, B., & Nagengast, B. (2009). Doubly-latent models of school contextual effects: Integrating multilevel and structural equation approaches to control measurement and sampling error. *Multivariate Behavioral Research, 44*(6), 764–802.
Marsh, H. W., & Parker, J. W. (1984). Determinants of student self-concept: Is it better to be a relatively large fish in a small pond even if you don't learn to swim as well? *Journal of Personality and Social Psychology, 47*(1), 213–231.
Mayer, A., Nagengast, B., Fletcher, J., & Steyer, R. (2014). Analyzing average and conditional effects with multigroup multilevel structural equation models. *Frontiers in Psychology, 5*, 1–16.
McCullagh, P., & Nelder, J. A. (1989). *Generalized linear models* (2nd ed.). London: Chapman and Hall.
McDonald, R. P., & Goldstein, H. (1989). Balanced versus unbalanced designs for linear structural relations in two-level data. *British Journal of Mathematical and Statistical Psychology, 42*(2), 215–232.
McNeish, D. M., & Stapleton, L. M. (2016). The effect of small sample size on two-level model estimates: A review and illustration. *Educational Psychology Review, 28*(2), 295–314.
Mehta, P. D. (2013a). n-level structural equation modeling. In Y. M. Petscher, C. Schatschneider, & D. L. Compton (Eds.), *Applied quantitative analysis in the social sciences* (pp. 329–361). New York: Routledge.

Mehta, P. D. (2013b). *N-level structural equation modeling: Xxm user's guide, version 1.0.* Retrieved from http://www2.gsu.edu/ wwwml1/wkshop/xxm.pdf

Mehta, P. D., & Neale, M. C. (2005). People are variables too: Multilevel structural equations modeling. *Psychological Methods, 10*(3), 259–284.

Meredith, W. (1993). Measurement invariance, factor analysis and factorial invariance. *Psychometrika, 58*(4), 525–543.

Miles, J., & Shevlin, M. (2007). A time and a place for incremental fit indices. *Personality and Individual Differences, 42*(5), 869–874.

Moineddin, R., Matheson, F. I., & Glazier, R. H. (2007). A simulation study of sample size for multilevel logistic regression models. *BMC Medical Research Methodology, 7*(1), 34–44.

Muthén, B. O. (1989). Latent variable modeling in heterogeneous populations. *Psychometrika, 54*(4), 557–585.

Muthén, B. O. (1994). Multilevel covariance structure analysis. *Sociological Methods & Research, 22*(3), 376–398.

Muthén, B. O. (2002). Beyond SEM: General latent variable modeling. *Behaviormetrika, 29*(1), 81–117.

Muthén, B. O., & Asparouhov, T. (2008). Growth mixture modeling: Analysis with non-Gaussian random effects. In G. Fitzmaurice, M. Davidian, G. Verbeke, & G. Molenberghs (Eds.), *Longitudinal data analysis* (pp. 143–166). Boca Raton, FL: CRC Press.

Muthén, B. O., & Asparouhov, T. (2012). Bayesian structural equation modeling: A more flexible representation of substantive theory. *Psychological Methods, 17*(3), 313–335.

Muthén, B. O., & Asparouhov, T. (2018). Recent methods for the study of measurement invariance with many groups: Alignment and random effects. *Sociological Methods & Research, 47*, 637–664.

Muthén, B. O., Khoo, S.-T., & Gustafsson, J.-E. (1997). *Multilevel latent variable modeling in multiple populations* (Technical Report). Los Angeles: Graduate School of Education & Information Studies, University of California.

Muthén, L. K., & Muthén, B. O. (1998–2017). *Mplus user's guide: Eighth edition.* Los Angeles, CA: Muthén & Muthén.

Nagengast, B., & Marsh, H. W. (2012). Big fish in little ponds aspire more: Mediation and cross-cultural generalizability of school-average ability effects on self-concept and career aspirations in science. *Journal of Educational Psychology, 104*(4), 1033–1053.

Nestler, S., & Back, M. D. (2017). Using cross-classified structural equation models to examine the accuracy of personality judgments. *Psychometrika, 82*(2), 475–497.

Paccagnella, O. (2006). Centering or not centering in multilevel models? The role of the group mean and the assessment of group effects. *Evaluation Review, 30*(1), 66–85.

Palardy, G. J. (2008). Differential school effects among low, middle, and high social class composition schools: A multiple group, multilevel latent growth curve analysis. *School Effectiveness and School Improvement: An International Journal of Research, Policy and Practice, 19*(1), 21–49.

Petty, R. E., & Cacioppo, J. T. (1986). *Communication and persuasion: Central and peripheral routes to attitude change.* New York: Springer-Verlag.

Pinheiro, J. C., & Bates, D. M. (2000). *Mixed-effects models in S and S-PLUS.* New York: Springer.

Preacher, K. J. (2011). Multilevel SEM strategies for evaluating mediation in three-level data. *Multivariate Behavioral Research, 46*(4), 691–731.

Preacher, K. J., Wichman, A. L., MacCallum, R. C., & Briggs, N. E. (2008). *Latent growth curve modeling*. Thousand Oaks, CA: SAGE.

Preacher, K. J., Zyphur, M. J., & Zhang, Z. (2010). A general multilevel SEM framework for assessing multilevel mediation. *Psychological Methods, 15*(3), 209–233.

Przeworski, A., Alvarez, M. E., Cheibub, J. A., & Limongi, F. (2000). *Democracy and development: Political institutions and well-being in the world, 1950–1990*. New York: Cambridge University Press.

Pugesek, B. H., Tomer, A., & Von Eye, A. (2003). *Structural equation modeling: Applications in ecological and evolutionary biology*. New York: Cambridge University Press.

Rabe-Hesketh, S., Skrondal, A., & Pickles, A. (2004). Generalized multilevel structural equation modeling. *Psychometrika, 69*(2), 167–190.

Raudenbush, S. W., & Bryk, A. S. (2002). *Hierarchical linear models: Applications and data analysis methods*. Thousand Oaks, CA: SAGE.

Raudenbush, S. W., & Liu, X. (2000). Statistical power and optimal design for multisite randomized trials. *Psychological Methods, 5*(2), 199–213.

Raykov, T., & Marcoulides, G. A. (2000). *A first course in structural equation modeling*. Mahwah, NJ: Lawrence Erlbaum.

Rhemtulla, M., Brosseau-Liard, P. E., & Savalei, V. (2012). When can categorical variables be treated as continuous? a comparison of robust continuous and categorical sem estimation methods under suboptimal conditions. *Psychological Methods, 17*(3), 354–373.

Rosseel, Y. (2012). lavaan: An R package for structural equation modeling. *Journal of Statistical Software, 48*(2), 1–36. Retrieved from http://www.jstatsoft.org/v48/i02/

Rubin, D. B. (1976). Inference and missing data. *Biometrika, 63*(3), 581–592.

Ryu, E., & Mehta, P. D. (2017). Multilevel factorial invariance in n-level structural equation modeling (nSEM). *Structural Equation Modeling: A Multidisciplinary Journal, 24*(6), 936–959.

Ryu, E., & West, S. G. (2009). Level-specific evaluation of model fit in multilevel structural equation modeling. *Structural Equation Modeling: A Multidisciplinary Journal, 16*(4), 583–601.

Schafer, J. L. (2001). Multiple imputation with PAN. In L. M. Collins & A. Sayer (Eds.), *New methods for the analysis of change* (pp. 357–377). Washington, DC: American Psychological Association.

Schafer, J. L., & Yucel, R. M. (2002). Computational strategies for multivariate linear mixed-effects models with missing values. *Journal of Computational and Graphical Statistics, 11*(2), 437–457.

Schumacker, R. E., & Lomax, R. G. (2004). *A beginner's guide to structural equation modeling* (2nd ed.). Mahwah, NJ: Lawrence Erlbaum.

Schwarz, G. (1978). Estimating the dimension of a model. *Annals of Statistics, 6*(2), 461–464.

Singer, J. D., & Willet, J. B. (2003). *Applied longitudinal data analysis*. Oxford, UK: Oxford University Press.

Snijders, T. A. B. (2005). Power and sample size in multilevel modeling. In B. S. Everitt & D. C. Howell (Eds.), *Encyclopedia of statistics in behavioral science: Vol. III* (pp. 1570–1573). Chichester, UK: John Wiley.

Snijders, T. A. B., & Bosker, R. J. (1993). Standard errors and sample sizes for two-level research. *Journal of Educational and Behavioral Statistics, 18*(3), 237–259.

Snijders, T. A. B., & Bosker, R. J. (1999). *Multilevel analysis: An introduction to basic and advanced multilevel modeling.* London: SAGE.

Snijders, T. A. B., & Bosker, R. J. (2012). *Multilevel analysis: An introduction to basic and advanced multilevel modeling* (2nd ed.). London: SAGE.

Sobel, M. E. (1982). Asymptotic confidence intervals for indirect effects in structural equation models. *Sociological Methodology, 13*, 290–312.

Sobel, M. E. (1986). Some new results on indirect effects and their standard errors in covariance structure models. *Sociological Methodology, 16*, 159–186.

Solt, F. (2008). Economic inequality and democratic political engagement. *American Journal of Political Science, 52*(1), 48–60.

Spiegelhalter, D. J., Best, N. G., Carlin, B. P., & van der Linde, A. (2002). Bayesian measures of model complexity and fit. *Journal of the Royal Statistical Society: Series B, 64*(4), 583–639.

Stapleton, L. M. (2002). The incorporation of sample weights into multilevel structural equation models. *Structural Equation Modeling: A Multidisciplinary Journal, 9*(4), 475–502.

Steele, F., Clarke, P., Leckie, G., Allan, J., & Johnston, D. (2017). Multilevel structural equation models for longitudinal data where predictors are measured more frequently than outcomes: An application to the effects of stress on the cognitive function of nurses. *Journal of the Royal Statistical Society: Series A (Statistics in Society), 180*(1), 263–283.

Stegmueller, D. (2013). How many countries for multilevel modeling? A comparison of frequentist and Bayesian approaches. *American Journal of Political Science, 57*(3), 748–761.

Tinnermann, A., Geuter, S., Sprenger, C., Finsterbusch, J., & Büchel, C. (2017). Interactions between brain and spinal cord mediate value effects in nocebo hyperalgesia. *Science, 358*(6359), 105–108.

van Buuren, S. (2011). Multiple imputation of multilevel data. In J. Hox & J. K. Roberts (Eds.), *Handbook of advanced multilevel analysis* (pp. 173–196). Routledge.

van de Vijver, F. J. R., & Poortinga, Y. H. (2002). Structural equivalence in multilevel research. *Journal of Cross-Cultural Psychology, 33*(2), 141–156.

Verhagen, A. J., & Fox, J. P. (2013). Bayesian tests of measurement invariance. *British Journal of Mathematical and Statistical Psychology, 66*(3), 383–401.

Welzel, C., & Inglehart, R. F. (2010). Agency, values, and well-being: A human development model. *Social Indicators Research, 97*(1), 43–63.

Wiley, D. E. (1973). The identification problem for structural equation models with unmeasured variables. In A. S. Goldberger & O. D. Duncan (Eds.), *Structural equation models in the social sciences* (pp. 69–83). New York: Seminar.

Wolfle, L. M. (2003). The introduction of path analysis to the social sciences, and some emergent themes: An annotated bibliography. *Structural Equation Modeling: A Multidisciplinary Journal, 10*(1), 1–34.

Yuan, K.-H., & Bentler, P. M. (2007). Multilevel covariance structure analysis by fitting multiple single-level models. *Sociological Methodology, 37*(1), 53–82.

Zhang, Z., Zyphur, M. J., & Preacher, K. J. (2009). Testing multilevel mediation using hierarchical linear models: Problems and solutions. *Organizational Research Methods, 12*(4), 695–719.

INDEX

Absolute badness-of-fit index, 12–13
Absolute model fit, 11–12
Akaike's information criterion (AIC), 9, 11
Arrows in path model, 5–6
Badness of fit, 10–14, 29. *See also* Model fit
Bayesian estimation, 18, 19
  in cross-classified data structures, 111
  in multilevel CFA with random loadings, 70, 75–76
Bayesian information criterion (BIC), 9, 11
Bayesian models, statistical significance in, 6
Bentler's comparative fit index (CFI), 11, 13
Between-level latent variables, in structural models, 88–94
Between-level structural model, 25, 26, 27–28
  centering in MSEM, 51–52
BIC (Bayesian information criterion), 9, 11
Big fish little pond effects (BFLPE), 86
Biology, genetic studies applications, 2–3
Bollen-Stine bootstrap, 8

Centering in MSEM, 51–52
CFA. *See* Confirmatory factor analysis
CFI (Bentler's comparative fit index), 11, 13
Chained equations, 113

Chi-square (χ^2) model fit, 10–12, 14
Clustered sampling design, 109–110
Clustered standard errors, 110
Coefficients in MLM, 15
Comparative fit index (CFI), Bentler's, 11, 13
Complementary log-log function, 108
Completely at random (MCAR), 9n
Configural constructs, 85
Configural invariance model, 57
Configurational construct, 52
Confirmatory factor analysis (CFA)
  in multiple groups, 57–58
  in SEM, 2, 3, 54, 56
Contextual variables, 85
Continuous outcomes, 108
Credibility intervals, 75
Cross-classified data structures, 19, 110–111
Cross-level interactions, 18
Cross-level invariance, 66, 89, 94

Degrees of freedom, 6–7
Design effect, 16
Deviance information criterion (DIC), 75–76
Deviance, in model fit, 9–10, 11
Disturbance (residual variances), 5
Dominican Republic example, 54, 69
Doubly latent model, 96

Education, in self-expression values example, 40–41
Effective sample size, in MLM, 16

123

Elaboration likelihood model, 49
Empirical underidentification, 7
Employee perceptions example, 78–79. *See also* Multilevel structural equation models
Endogenous variables, 2, 20
Equality constraints, 67, 89, 94, 96
Errors (residual variances), 5, 7
Estimation, simultaneous, 7
Exogenous variables, 2, 20

Factor-analytic approaches, 55
Factor loading estimates, statistical significance in, 6
Factor loading notation in MSEM, 20
Factor loadings, 5, 7
 in full structural model in multilevel setting, 26
Factor models. *See* Multilevel factor models
Factors. *See* Latent variables
Fit measures. *See* Model fit
Fixed effects, 17, 19, 110
Fixing factor loadings and residual variances, 7
Free parameters, 61
Frequentist confidence intervals, 75n
Full-information maximum likelihood (FIML), 9n, 18, 114
Full structural equation model, 3
Full structural models in multilevel setting, 24–29
Fully conditional specification, 113

GDP, natural logarithm of, 34
Generalized least squares (GLS) estimation, 18n
Generalized linear latent and mixed modeling (GLLAMM), 109
Generalized linear modeling (GLM), 108

Goodness of fit, 9, 13. *See also* Model fit
Grand mean, 59
Grand mean centering, 51
Graphical notation
 for MLM, 15 (figure)
 for SEM, 4 (figure)
 See also Notation
Group mean centering, 51
Group membership, 114
Growth model, 106

Household surveys, 19n
Hyperparameters, 17

Identification
 in multilevel CFA, 61–62
 in SEM, 6–7, 80
Identity link function, 108
Imputation and missing data, 113
Independence model, 60
Intercepts for variables
 formula specifications of, 3
 in random intercepts model, 22–23
Intraclass correlation coefficients (ICCs), 60–61
Invariant CFA models, 57, 66

JKW model, 2
Joint modeling, 113
Joreskog, Karl, 2
Just-identified models, 6, 12

Keesling, Ward, 2

Latent contextual covariate model, 85–88
Latent growth curve, 106–107
Latent variable models, 3
Latent variables
 between-level, structural models with, 88–98
 multilevel measurement models with, 54, 57, 76
 in multilevel settings, 22, 26–28

random intercepts model with
random mean, 96–97
in random slopes model, 50
random slopes model with
random mean, 98–102
in SEM, 2
lavaan software package, 30
Least squares-based methods, 9
Levels of analysis
in clustered sampling design, 109–110
in mediation, 48, 49
in MLM, 15–19
in random latent variable intercepts, 66–67
Likelihood, 8
Likelihood ratio test, 9–11, 18
Limit of close fit, 13
Linear additive modeling specification, 108
Linear predictor, 108
Listwise deletion, 10, 34, 54
Logit function, 108
Log-likelihood (LL), 8
Longitudinal studies, 19n

Matrix algebra, 20, 28
Maximum likelihood, 8
Maximum likelihood estimation, 8
multilevel models, 18, 19
Maximum likelihood fit function (F_{ML}), 12
Mean structure, 3, 5
Measurement invariance
in multilevel CFA with random loadings, 71–72
and noninvariance, in CFA in multilevel groups, 57
Mediation in multilevel path models, 39, 45–48
Mediator variable, 45
Metric invariance model, 57
Micro-macro effects, 46
Missing at random (MAR), 9n
Missing data, 113–114

Mixed-effects model, 15, 17. *See also* Multilevel modeling
MLM. *See* Multilevel modeling
MLR (robust maximum likelihood), 8
Model comparison
of MSEM models, 29
of multilevel CFA with random loadings, 75–76
of multilevel structural equation models, 87, 94, 98, 106–117, 114–115
of random intercepts and random slopes models, 45
of SEM and traditional regression models, 9–14
Model fit
of MSEM, 29–30, 104–107
of multilevel factor models, 75–76
of multilevel models, 18
of multilevel path models, 45
of SEM, 9–14, 114–115
Model-implied variance-covariance matrix, and simultaneous estimation, 7–8
Moderation in multilevel path models, 45, 48–51
Mplus software
availability of, 30
centering in MSEM, 52
number of variables in demo version, 103n
three-level models, 107n
MSEM. *See* Multilevel structured equation modeling
Multicollinearity, 64
Multigroup analysis, in SEM, 112
Multigroup multilevel CFA, 56n
Multigroup SEM, 57n
Multilevel CFA, 58
two-level CFA. *See* Two-level CFA
Multilevel CFA with random loadings, 69–76
example, 72–75

measurement invariance, 71–72
model fit and comparison, 75–76
Multilevel exploratory factor
  analysis (EFA), 54n
Multilevel factor models, 54–77
  CFA in multiple groups, 57–58
  multilevel CFA with random
    loadings, 69–76
  random latent variable
    intercepts, 66–69
  summary, 76–77
  two-level CFA. *See* Two-level
    CFA
Multilevel imputation model, 113
Multilevel latent covariate model,
  85–88
Multilevel modeling (MLM), 15–20
  combining and compared with
    SEM, 1, 106–107, 114–115
  estimation and model fit, 18,
    114–115
  further readings, 19–20
  imputation and missing data, 113
  notation conventions, 15–17
  sample size, 16, 18–19
Multilevel path models, 20–24, 31–53
  about path models, 31–32
  centering in MSEM, 51–52
  mediation, 45–48, 51
  moderation, 45, 48–51
  multilevel regression example,
    34–36
  notation, 20–24
  random intercepts model, 36–41, 45
  random slopes model, 41–45
  summary, 52–53
Multilevel regression example, 34–36
Multilevel structural equation
  models, 78–105
  with between-level latent
    variables, 88–98
  bringing factor and path models
    together, 78–79

model comparison. *See* Model
  comparison
multilevel latent covariate
  model, 85–88
random intercept of observed
  outcome, 80–84
random intercepts of latent
  variables, 94–98
random slopes MSEM, 98–102
summary, 105
Multilevel structured equation
  modeling (MSEM)
  estimation and model fit, 29–30
  examples of. *See* Multilevel
    structural equation models
  full structural models in
    multilevel setting, 24–29
  introduction to, 20–30
  multilevel path model, 20–24
  notation and graphical form,
    20–23, 25
  online materials, 30
Multiple group analysis, 57
Multiple-group CFA (MGCFA),
  57–58
  invariance testing, 71–72
Multiple-membership models, 19,
  111
Multitrait multimethod (MTMM),
  112

Nested models, 10–11
N-level structural equation
  modeling (NL-SEM, nSEM),
  107, 111
Nonhierarchical data structures,
  110–111
Noninvariant CFA models, 57, 72
Nonredundant elements,
  calculating number of, 6
Notation
  for MLM, 15–17
  for MSEM, 20–24
  for SEM, 3–6

nSEM. *See* N-level structural equation modeling (NL-SEM)
Null model, 13

Observations, formula specifications of, 3
Observed outcome, random intercept model of, 80–84
Observed variance-covariance matrix, 6–8
Occupational hierarchy example, 31
Ordinary least squares (OLS) regression, 16–17
Overidentified models, 6, 7, 61, 80

Partially saturated model fit test, 29–30, 63–64
Path analysis models, 2
  with latent variables, example of, 4 (figure)
  See also Multilevel path models
Path models, multilevel, 20
Political science applications, 15
Predictor/predict terms, 32
Probit function, 108
Product-of-coefficients method, 47
Professional aspirations example, 31–32
Programme for International Student Assessment (PISA), 54–57, 60, 69, 76, 112
Psychology applications, 2, 7

Random coefficients specifications, 15. *See also* Multilevel modeling (MLM)
Random effects, 17, 18
Random intercepts model, 22–23
  with between-level latent variables, 88–94, 95
  in multilevel modeling, 58
  as multilevel path example, 36–41, 43–45
  of observed outcome, 80–84, 102–104

with random mean latent variables, 96–98
Random intercepts random slopes model, 17, 25–29
Random latent variable intercepts, 66–69
Random loadings models, 70, 72
Random parameters, 17
Random slopes model
  and multilevel CFA with random loadings, 70
  as multilevel path example, 41–45, 50
  with random mean latent variables, 98–102, 103–104
Regression coefficients in MSEM, 20
Relative power theory, 49
Residuals notation, 3, 20
Residual variances, 5, 7
Restricted maximum likelihood (REML), 18, 19
Robust maximum likelihood (MLR), 8
Root mean square error of approximation (RMSEA), 11, 13

Sample size
  in MLM, 16, 18–19
  in MSEM, clusters and parameters, 105
Satorra-Bentler adjusted χ^2, 13
Satorra-Bentler scaled χ^2, 13
Saturated, partially, model fit test, 29–30, 63–64
Scalar invariance model, 57
Seemingly unrelated regression, 54
Self-expression values example, 32–34. *See also* Multilevel path models
SEM. *See* Structural equation modeling
Simpson's paradox, 98
Simultaneous estimation, 7–8

Single-equation estimation, 9
Slopes
  formula specifications of, 3
  in random intercepts model, 22–23
  See also Random slopes model
Social sciences, 1–2, 49, 108, 109–110
Software. *See* lavaan software package; Mplus software
Standardized root mean square residual (SRMR), 11, 14, 29–30
Statistical significance, 5–6
Strict invariance model, 57
Structural equation modeling (SEM)
  combined and compared with MLM, 1, 106–107, 114–115
  defined, 2
  for direct and indirect statistical associations, 36
  estimation, 7–9
  full model in multilevel setting, 24–29
  further readings, 14–15, 107
  identification, 6–7
  model fit, 9–14, 114–115
  model specification and notation, 3–6
Structural equation modeling, n-level (NL-SEM), 107, 111
Student Internet usage example, 55–57. *See also* Multilevel factor models
Student performance example, 54
Superscripts in MSEM, 20–21

Three-stage least squares (3SLS), 9
Time-series cross-sectional models, 113n
Tracing rules, 47
Transmission mechanism, 46
Two-level CFA, 58–66
  estimation, 60–61
  identification, 61–62
  partial saturation fit test, 29–30, 63–64
  results, 62–63
  unidimensional level 2 factor structure, 64–66
Two-level MSEM with random intercept, 80–84
Two-stage least squares (2SLS), 9

Underidentified models, 6
Unidimensional level 2 factor structure, 64–66

Variance-covariance matrix, 11–12
  estimating cells in, 32
  model-implied, 7–8
  observed, 6–8
Vectors in matrices, 28

Website, 30
Weighting, 56n, 109
Wiley, David, 2
Within-level structural model, 25, 27
  centering in MSEM, 51–52
Within-person observation design, 106
Workplace Employment Relations Survey (WERS), 78–79
World Values Surveys (WVS), 32–34, 35
Wright, Sewell, 2

Milton Keynes UK
Ingram Content Group UK Ltd.
UKHW020043120424
441015UK00011B/432